Topics in Geometry, Coding Theory and Cryptography

T0214876

Algebra and Applications

Volume 6

Managing Editor:

Alain Verschoren
RUCA, Belgium

Series Editors:

Christoph Schweigert
Hamburg University, Germany

Ieke Moerdijk
Utrecht University, The Netherlands

John Greenlees
Sheffield University, UK

Mina Teicher
Bar-Ilan University, Israel

Eric Friedlander
Northwestern University, USA

Idun Reiten
Norwegian University of Science and Technology, Norway

Algebra and Applications aims to publish well written and carefully refereed monographs with up-to-date information about progress in all fields of algebra, its classical impact on commutative and noncommutative algebraic and differential geometry, K-theory and algebraic topology, as well as applications in related domains, such as number theory, homotopy and (co)homology theory, physics and discrete mathematics.

Particular emphasis will be put on state-of-the-art topics such as rings of differential operators, Lie algebras and super-algebras, group rings and algebras, C*algebras, Kac-Moody theory, arithmetic algebraic geometry, Hopf algebras and quantum groups, as well as their applications. In addition, Algebra and Applications will also publish monographs dedicated to computational aspects of these topics as well as algebraic and geometric methods in computer science.

Topics in Geometry, Coding Theory and Cryptography

Edited by

Arnaldo Garcia

Instituto de Matematica Pura e Aplicada (IMPA),
Rio de Janeiro, Brazil

and

Henning Stichtenoth

University of Duisburg-Essen, Germany and
Sabanci University, Istanbul, Turkey

 Springer

A C.I.P. Catalogue record for this book is available from the Library of Congress.

ISBN-13 978-90-481-7345-7
ISBN-10 1-4020-5334-4 (e-book)
ISBN-13 978-1-4020-5334-4 (e-book)

Published by Springer,
P.O. Box 17, 3300 AA Dordrecht, The Netherlands.

www.springer.com

Printed on acid-free paper

Contents

Foreword

The theory of algebraic function fields has a long history. Its origins are in number theory, and there are close interrelations with other branches of pure mathematics such as algebraic geometry or compact Riemann surfaces. In fact, the study of algebraic function fields is essentially equivalent to the study of algebraic curves. These relations have been well-known for a long time.

Around 1980 V. D. Goppa came up with a brilliant idea of constructing error-correcting codes by means of algebraic function fields over finite fields. These codes are now known as *geometric Goppa codes* or *algebraic geometry codes* (AG *codes*). The key point of Goppa's construction is that one gets information about the code parameters (length, dimension, minimum distance of the code) in terms of geometric and arithmetic data of the function field (number of rational places, genus). Goppa's method can be seen as a "simple" generalization of the construction of Reed-Solomon codes: one just replaces the evaluation of polynomials in one variable at elements of a finite field (which is used for the definition of Reed-Solomon codes) by evaluating functions of a function field at some of its rational places. A basic role is then played by the Riemann-Roch theorem.

Soon after Goppa's discovery, M. A. Tsfasman, S. G. Vladut and T. Zink constructed families of AG codes of increasing length whose asymptotic parameters are better than those of all previously known infinite sequences of codes and which beat the Gilbert-Varshamov bound - a bound which is well-known in coding theory and which is a classical measure for the performance of long codes. The proof of the Tsfasman-Vladut-Zink result uses two main tools: Goppa's construction of AG codes and the existence of curves or function fields (more specifically: classical or Drinfeld modular curves) over a finite field having large genus and many rational places.

Cyclic codes have a natural representation as *trace codes*, and one can associate with each codeword of a trace code an Artin-Schreier function field. Properties of this function field (specifically the number of rational places) reflect properties of the corresponding cyclic code (namely the weights of codewords and subcodes). In this way one gets another link between codes and function fields which is entirely different from Goppa's.

In 1985, N. Koblitz invented cryptosystems which are based on elliptic curves (or elliptic function fields) over a finite field. These cryptosystems are very powerful and attracted much attention; they created a new and very lively area of research (*elliptic curve cryptography*) and brought together researchers from pure mathematics (number theory, arithmetic geometry) and applied mathematics and engineering (cryptography). Similar as in the case of coding theory, this interaction proved fruitful for both sides, posing new problems and leading to many interesting practical and theoretical results.

The above-mentioned applications of function fields in constructing good long codes (due to Goppa and to Tsfasman-Vladut-Zink) and in constructing powerful cryptosystems via elliptic or hyperelliptic curves are now well-known. However, most mathematicians and engineers are not so familiar with many other, entirely different applications of function fields. To mention some of them: dense *sphere packings* in high-dimensional spaces; sequences with *low discrepancy*; *multiplication algorithms* in finite fields; the construction of *nonlinear codes* whose asymptotic parameters are even better than the Tsfasman-Vladut-Zink bound; the construction of good *hash families*. In all these cases the use of function fields leads to better results than those of classical approaches.

In this book we present five survey articles on some of these new developments. Most of the material is directly related to the interactions between function fields and their various applications; in particular the structure and the number of rational places of function fields are always of great significance. When choosing the topics, we also tried to focus on material which has not yet been presented in books or review articles. So, for instance, we did not include chapters about elliptic curve cryptography or about AG codes. There are numerous interconnections between the individual articles. Wherever applications are pointed out, a special effort has been made to present some background concerning their use. For the convenience of the reader, we have included an appendix which summarizes the basic definitions and results from the theory of algebraic function fields.

We give now a brief summary of the five chapters. More detailed descriptions are given in the introduction of each chapter.

Chapter 1. *Towers of Algebraic Function Fields over Finite Fields, by Arnaldo Garcia and Henning Stichtenoth.* In this chapter, the authors give a comprehensive survey of their work on explicit towers of algebraic function fields having many rational places. This concept provides a more elementary and explicit approach than class field towers and towers from modular curves. Towers with many rational places play a crucial role in many "asymptotic" constructions, such as error-correcting codes (Tsfasman-Vladut-Zink), low-discrepancy sequences (Niederreiter-Xing), and other applications of function fields in cryptography (see Chapter 2). Several examples of asymptotically good recursive towers are presented in detail. The proofs for the behaviour of the genus in wild towers are considerably simplified, compared to the proofs in the original papers.

Chapter 2. *Function Fields over Finite Fields and Their Applications to Cryptography, by Harald Niederreiter, Huaxiong Wang and Chaoping Xing.* This survey article focuses on several recent, less well-known applications of function fields – specifically, function fields with many rational places – in cryptography and combinatorics. Many of these applications are due to the authors. Among the topics are constructions of authentication codes, frameproof codes, perfect hash families, cover-free families and pseudorandom sequences of high linear complexity.

Chapter 3. *Artin-Schreier Extensions and Their Applications, by Cem Güneri and Ferruh Özbudak.* Extensions of function fields of Artin-Schreier type provide many examples of function fields having many rational places; this makes them very interesting for coding theory. In this chapter, several other applications of Artin-Schreier extensions are discussed, among them to the famous Weil bound for character sums, to weights of trace codes and to generalizations of cyclic codes.

Chapter 4. *Pseudorandom Sequences, by Alev Topuzoğlu and Arne Winterhof.* Various constructions of pseudorandom sequences are based on function fields, see Chapters 2 and 5. Therefore, some background material on the theory of pseudorandom sequences is presented in Chapter 4. In particular, the important concept of linear complexity and some related measures for the performance of pseudorandom sequences are discussed in this chapter.

Chapter 5. *Group Structure of Elliptic Curves over Finite Fields and Applications, by Ram Murty and Igor Shparlinski.* Motivated by applications of elliptic curves to cryptography, the structure of the group of $\mathbb{F}_q$-rational points of an elliptic curve has attracted much attention. In particular it is an important feature for cryptographic applications if this group is cyclic or if it contains a large cyclic subgroup. The authors give a survey of recent results on this topic. Techniques from many branches of number theory and algebraic geometry are used in this chapter.

Each chapter begins with a detailed introduction, giving an overview of its contents and also giving some applications and motivation. It is clear that we do not want to present all proofs here. However, whenever possible, some typical proofs are provided. Our aim is to stimulate further research on some promising topics at the border line between pure and applied mathematics; therefore each chapter contains also an extensive list of references of recent research papers.

Some of the authors (A. Garcia, H. Niederreiter, I. Shparlinski, H. Stichtenoth, A. Winterhof and C. Xing) visited Sabancı University in Istanbul (Turkey) during the years 2002-2005, where they presented part of the material of this volume. It is our pleasure to thank our hosts at Sabancı University for their support and hospitality.

January 2006

Arnaldo Garcia, Henning Stichtenoth

Chapter 1

EXPLICIT TOWERS OF FUNCTION FIELDS
OVER FINITE FIELDS

Arnaldo Garcia and Henning Stichtenoth

1.   Introduction

The purpose of this review article is to serve as an introduction and at the same time, as an invitation to the theory of towers of function fields over finite fields. More specifically, we treat here the case of explicit towers; i.e., towers where the function fields are given by explicit equations. The asymptotic behaviour of the genus and of the number of rational places in towers are important features for applications to coding theory and to cryptography (cf. Chapter 2).

The interest in solutions of algebraic equations over finite fields has a long history in mathematics, especially when the equations define a one-dimensional object (a curve or, equivalently, a function field). The major result of this theory is the Hasse-Weil theorem which gives in particular an upper bound for the number of rational points in terms of the genus of the curve and of the cardinality of the finite field.

The Hasse-Weil theorem is equivalent to the validity of Riemann's Hypothesis for the Zeta function associated to the curve by E. Artin, in analogy with the classical situation in Number Theory. This upper bound of Hasse-Weil is sharp, and the curves attaining this bound are called maximal curves. Y. Ihara was the first to notice that the Hasse-Weil bound can be improved for curves of high genus, and he gave in particular an upper bound for the genus of maximal curves in terms of the cardinality of the finite field.

We will use here the language of function fields; i.e., we will be closer to Number Theory than to Algebraic Geometry. Hence the concepts we will deal with are function fields, field extensions, traces, norms, valuations, places, ratio-

A. Garcia and H. Stichtenoth (eds.), Topics in Geometry, Coding Theory and Cryptography, 1–58.
© 2007 *Springer.*

nal places, ramification indices and inertia degrees, tame and wild ramification, etc.

Denote by $\mathbb{F}_q$ the finite field of cardinality q. For a function field F over $\mathbb{F}_q$ we denote by $N(F)$ its number of $\mathbb{F}_q$-rational places and by $g(F)$ its genus. The upper bound of Hasse-Weil is

$$N(F) \leq 1 + q + 2\sqrt{q} \cdot g(F),$$

and Ihara showed that if the equality holds above then $2g(F) \leq q(q-1)$.

The following real number

$$A(q) := \limsup_{g(F) \to \infty} N(F)/g(F),$$

where F runs over all function fields over the field $\mathbb{F}_q$, was introduced by Ihara. It is of fundamental importance for the theory of function fields over a finite field, since it gives information about how many rational places a function field $F/\mathbb{F}_q$ of large genus can have.

In order to investigate the quantity $A(q)$, it is natural to study towers of function fields over $\mathbb{F}_q$; i.e., one considers sequences $\mathcal{F} = (F_0, F_1, F_2, \ldots)$ of function fields F_i over $\mathbb{F}_q$ with $F_0 \subseteq F_1 \subseteq F_2 \subseteq \ldots$ with the property $g(F_i) \to \infty$. It can be seen easily that the limit of the tower

$$\lambda(\mathcal{F}) := \lim_{n \to \infty} N(F_n)/g(F_n)$$

always exists (see Section 3), and it is clear that the estimate below holds:

$$0 \leq \lambda(\mathcal{F}) \leq A(q).$$

As follows from the Hasse-Weil bound, we have that $A(q) \leq 2\sqrt{q}$. Based on Ihara's ideas, this bound was improved by Drinfeld-Vladut who showed that

$$A(q) \leq \sqrt{q} - 1.$$

But even before this bound of Drinfeld-Vladut was obtained, Ihara (and independently Tsfasman-Vladut-Zink) proved that if q is a square then $A(q) \geq \sqrt{q} - 1$. We thus have the equality

$$A(q) = \sqrt{q} - 1, \quad \text{if } q \text{ is a square}.$$

The proofs given by Ihara and Tsfasman-Vladut-Zink use the fact that certain modular curves have many rational points. However these curves are in general not easy to describe by explicit equations. Another approach due to J.-P. Serre uses class field theory in order to prove the existence of curves of arbitrary high genus with sufficiently many rational points. Also this construction is not explicit. Our purpose here is to stimulate the investigation of explicit towers of function fields over finite fields; i.e., the function fields of the towers should be given explicitly by algebraic equations. The concept of explicit towers was first introduced in 1995 in the paper [20].

These notes are organized as follows:

- Section 2 contains basic concepts such as towers of function fields and their limits; recursive towers and the corresponding pyramids; tame and wild ramification in towers; linear codes and their parameters. In Section 2 one also finds:

  - The statement of the fundamental Hasse-Weil theorem (Theorem 2.3).
  - Serre's "explicit formulae" for bounding the number of rational places in a function field (Proposition 2.4).
  - The Drinfeld-Vladut bound (Theorem 2.5).
  - The Tsfasman-Vladut-Zink theorem connecting the asymptotics of function fields with the asymptotics of linear codes (Theorem 2.7).
  - Abhyankar's lemma which is an important tool to study the behaviour of the genus in recursive towers (Theorem 2.11).

- Section 3 is devoted to the investigation of the behaviour of the genus and of the number of rational places in towers of function fields over finite fields. It contains the following notions: the genus and the splitting rate of a tower; subtowers; asymptotically good and asymptotically optimal towers; ramification locus and splitting locus of a tower. In Section 2 one also finds:

  - A proof that the limit of a tower exists (Definition 3.4).
  - The limit of a subtower is at least as big as the limit of the tower (Proposition 3.6).
  - A sufficient condition which ensures that the genus of a tower is finite (Theorem 3.8 and Corollary 3.9).
  - A sufficient condition which ensures that a tower has finite ramification locus (Proposition 3.10).
  - A sufficient condition which ensures the existence of completely splitting places (Proposition 3.13).
  - A sufficient condition which ensures that a polynomial $f(X, Y)$ does define a recursive tower (Proposition 3.14).

- In Section 4 we investigate some interesting recursive tame towers, in which every step F_{n+1}/F_n is a Kummer extension. It contains the following subsections:

- Section 4.1: The optimal tower T_1 over $\mathbb{F}_4$ which is given recursively by the equation $Y^3 = X^3/(X^2 + X + 1)$.
- Section 4.2: For $r \geq 2$ and $q = \ell^r$, the tower T_2 over $\mathbb{F}_q$ which is defined recursively by the equation

$$Y^m = (X + 1)^m - 1 \quad \text{with} \quad m = (q - 1)/(\ell - 1).$$

  This tower gives a very simple proof that $A(q) > 0$ if q is not a prime number.

- Section 4.3: For $q = p^2$ and p an odd prime number, the optimal tower T_3 over $\mathbb{F}_q$ given recursively by the equation

$$Y^2 = (X^2 + 1)/2X.$$

  This tower corresponds to the modular curves $X_0(2^n)$ and it reveals some remarkable properties of Deuring's polynomial.

We also mention in Section 4 some other interesting towers from the papers [3, 14, 24, 33, 37].

- Section 5 is devoted to recursive wild towers. Especially interesting are wild towers where every step F_{n+1}/F_n is an Artin-Schreier extension, since some of the best towers known in the literature are of this type. We present here a simple method which allows a unified treatment of the genus behaviour of several towers of Artin-Schreier type (Lemma 5.1). Section 5 contains the following subsections:

  - Section 5.1: The optimal tower W_1 over $\mathbb{F}_q$ with $q = \ell^2$, which is defined recursively by the equation

$$Y^\ell + Y = X^\ell/(X^{\ell-1} + 1).$$

    A complete proof for the optimality of the tower W_1 is given and this proof is much simpler than the original one in [21].

  - Section 5.2: The optimal tower W_2 over $\mathbb{F}_q$ with $q = \ell^2$, which is the first explicit example in the literature attaining the Drinfeld-Vladut bound [20].

  - Section 5.3: The optimal tower W_3 over $\mathbb{F}_q$ with $q = \ell^2$, which is given recursively by the equation

$$(Y - 1)/Y^\ell = (X^\ell - 1)/X.$$

  - Section 5.4: The tower W_4 over the field with eight elements, which is recursively given by

$$Y^2 + Y = X + 1 + 1/X.$$

This tower was first introduced in [30], and we give here a much simpler proof for its asymptotic behaviour.

- Section 5.5: The tower $\mathcal{W}_5$ over the cubic field $\mathbb{F}_q$ with $q = \ell^3$ which is defined recursively by the equation

$$Y^\ell - Y^{\ell-1} = 1 - X - X^{-(\ell-1)}.$$

The tower $\mathcal{W}_5$ generalizes the tower $\mathcal{W}_4$ of Section 5.4, and its limit $\lambda(\mathcal{W}_5) \geq 2(\ell^2 - 1)/(\ell + 2)$ gives the best known lower bound for Ihara's quantity $A(\ell^3)$.

■ Section 6 contains some miscellaneous results on towers, among them a couple of conditions which easily show sometimes that a given tower is asymptotically bad (Theorem 6.2, Theorem 6.3 and Theorem 6.6). This section has the following subsections:

- Section 6.1: In a tower $(F_0, F_1, F_2, \ldots)$ of function fields, the growth of the genus $g(F_n)$ depends on the behaviour of the different degrees of the extensions F_n/F_{n-1}. This interrelation is explored in Theorem 6.1 and Theorem 6.2 where sufficient conditions are given for the tower to have finite or infinite genus.

- Section 6.2: Skew towers are asymptotically bad. This means: if the equation $f(X, Y) = 0$ which defines a recursive tower has unequal degrees in the variables X and Y, then the tower is asymptotically bad (Theorem 6.3).

- Section 6.3: Here the concept of the dual tower of a recursive tower is introduced; if the ramification loci of the tower and of its dual tower are distinct, then the tower is bad (Theorem 6.6).

- Section 6.4: This subsection contains a classification result on recursive towers defined by an Artin-Schreier equation of prime degree p of the form

$$Y^p + aY = \psi(X),$$

with $a \in \mathbb{F}_q^\times$ and with a rational function $\psi(X) \in \mathbb{F}_q(X)$. If such a tower is asymptotically good, then the function $\psi(X)$ must have a very specific form (Theorem 6.8).

2.   Towers and Codes

Throughout this Chapter we denote by $\mathbb{F}_q$ the finite field with q elements and by $p = \mathrm{char}(\mathbb{F}_q)$ its characteristic. We are interested in function fields over $\mathbb{F}_q$ (briefly, $\mathbb{F}_q$-function fields) having many rational places with respect to the genus. For basic concepts and facts about algebraic function fields (such as the definitions of function fields, places, divisors, rational places, genus, ramification, and Riemann-Roch theorem, Hurwitz genus formula, etc.) we refer to the Appendix or to [48]. For an $\mathbb{F}_q$-function field F we always assume throughout that $\mathbb{F}_q$ is the full constant field of F; i.e., that $\mathbb{F}_q$ is algebraically closed in F.

We denote by $N(F)$ the number of rational places and by $g(F)$ the genus of an $\mathbb{F}_q$-function field F, and we will be mainly interested in the behaviour of the ratio $N(F)/g(F)$ for function fields of large genus. To investigate this behaviour, Ihara [31] introduced the following quantity $A(q)$:

$$A(q) = \limsup_{g(F)\to\infty} N(F)/g(F),$$

where F runs over all function fields over $\mathbb{F}_q$. To deal with this quantity $A(q)$ one is naturally led to towers of function fields.

Definition 2.1. A *tower* $\mathcal{F}$ over $\mathbb{F}_q$ (or an $\mathbb{F}_q$-*tower*) is an infinite sequence $\mathcal{F} = (F_0, F_1, F_2, ...)$ of function fields $F_i/\mathbb{F}_q$ such that

i) $F_0 \subsetneqq F_1 \subsetneqq F_2 \subsetneqq \cdots \subsetneqq F_n \subsetneqq \cdots$;

ii) each extension F_{n+1}/F_n is finite and separable;

iii) the genera satisfy $g(F_n) \to \infty$ as $n \to \infty$.

For an $\mathbb{F}_q$-tower $\mathcal{F}$ the following limit does exist (see Section 3):

$$\lambda(\mathcal{F}) = \lim_{n\to\infty} N(F_n)/g(F_n).$$

It is clear from the definitions that one has

$$0 \le \lambda(\mathcal{F}) \le A(q).$$

Definition 2.2. The real number $\lambda(\mathcal{F})$ is called the *limit* of the $\mathbb{F}_q$-tower $\mathcal{F}$. The tower $\mathcal{F}$ is called *asymptotically good* if it has a positive limit $\lambda(\mathcal{F}) > 0$. If $\lambda(\mathcal{F}) = 0$ then $\mathcal{F}$ is said to be *asymptotically bad*.

It is not easy in general to construct asymptotically good towers, and it is an even harder task to construct towers over finite fields with large limits. These are the main concerns of this Chapter.

We start by deriving an upper bound for $A(q)$, the so-called Drinfeld-Vladut bound. It states that

$$A(q) \le \sqrt{q} - 1. \tag{2.1}$$

This bound is then also an upper bound for the limit of towers; i.e., the following inequality holds for all $\mathbb{F}_q$-towers $\mathcal{F}$:

$$\lambda(\mathcal{F}) \le \sqrt{q} - 1.$$

In order to prove the upper bound in (2.1) for $A(q)$ we will need the following theorem due to Hasse and Weil, which is the central result of the theory of function fields over finite fields. It is equivalent to the validity of the Riemann Hypothesis in this context, cf. [48, p.169]. Hasse [29] proved it for elliptic function fields (i.e., for $g(F) = 1$), and Weil [61] proved it in the general case. For other proofs of the Hasse-Weil theorem we refer to [10] and [47].

We need some notation: for a function field $F/\mathbb{F}_q$, let $F^{(r)} := F \cdot \mathbb{F}_{q^r}$ be the constant field extension of F of degree r, and let $N_r(F) := N(F^{(r)})$ be the number of $\mathbb{F}_{q^r}$-rational places of the function field $F^{(r)}$ over $\mathbb{F}_{q^r}$. The Hasse-Weil theorem can be stated in the following form:

Theorem 2.3. (Hasse-Weil) *Let F be an $\mathbb{F}_q$-function field of genus $g(F) = g$. Then there exist complex numbers $\alpha_1, \alpha_2, \ldots, \alpha_{2g} \in \mathbb{C}$ with the following properties:*

i) They can be ordered in such a way that

$$\alpha_{g+i} = \bar{\alpha}_i \text{ for } i = 1, \ldots, g.$$

ii) The polynomial $L(t) := \prod_{i=1}^{2g}(1 - \alpha_i t)$ has integer coefficients. It follows in particular that each α_i is an algebraic integer.

iii) For all $r \geq 1$ we have

$$N_r(F) = q^r + 1 - \sum_{i=1}^{2g} \alpha_i^r.$$

iv) The absolute value of α_i is

$$|\alpha_i| = \sqrt{q} \text{ for } i = 1, \ldots, 2g.$$

The elements $\alpha_i^{-1} \in \mathbb{C}$ are the roots of the Zeta function associated to the function field $F/\mathbb{F}_q$. From item iv) and item iii) with $r = 1$, one gets the so-called Hasse-Weil bound

$$N(F) \leq q + 1 + 2\sqrt{q} \cdot g(F).$$

This bound implies immediately that $A(q) \leq 2\sqrt{q}$. For the proof of the Drinfeld-Vladut bound (2.1) we make use of Serre's "explicit formulae":

Proposition 2.4. (Serre) (see [49]). *Let $0 \neq h(X) \in \mathbb{R}[X]$ be a polynomial with non-negative coefficients and with $h(0) = 0$. Suppose that the associated rational function $H(X)$, which is defined as*

$$H(X) = 1 + h(X) + h(X^{-1}),$$

satisfies the condition

$$H(\beta) \geq 0 \text{ for all } \beta \in \mathbb{C} \text{ with } |\beta| = 1.$$

Then for any function field $F/\mathbb{F}_q$ we have

$$N(F) \leq 1 + \frac{h(q^{1/2})}{h(q^{-1/2})} + \frac{g(F)}{h(q^{-1/2})}.$$

Proof. Let F be a function field over $\mathbb{F}_q$ with $g(F) = g$, and let $\alpha_1, \alpha_2, \ldots, \alpha_{2g}$ be the associated complex numbers, ordered as in item i) of Theorem 2.3. For simplicity we set $N_r(F) = N_r$ and in particular $N(F) = N_1$. Write

$$h(X) = \sum_{r=1}^{m} c_r X^r,$$

with $c_r \in \mathbb{R}$ and $c_r \geq 0$ for all r. Then we have

$$N_r = 1 + q^r - \sum_{i=1}^{g} (\alpha_i^r + \bar{\alpha}_i^r),$$

by item iii) of Theorem 2.3; hence

$$N_r \cdot q^{-r/2} = q^{-r/2} + q^{r/2} - \sum_{i=1}^{g} ((\alpha_i q^{-1/2})^r + (\bar{\alpha}_i q^{-1/2})^r)$$

$$(2.2)$$

$$= q^{-r/2} + q^{r/2} - \sum_{i=1}^{g} (\beta_i^r + \bar{\beta}_i^r),$$

with $\beta_i = \alpha_i q^{-1/2}$. By item iv) of Theorem 2.3, the complex numbers β_i have absolute value $|\beta_i| = 1$, so $\bar{\beta}_i = \beta_i^{-1}$. We now multiply Equation (2.2) by the coefficient c_r of $h(X)$ and we sum up for $r = 1, \ldots, m$, to obtain

$$\sum_{r=1}^{m} N_r \cdot c_r \cdot q^{-r/2} = h(q^{-1/2}) + h(q^{1/2}) + g - \sum_{i=1}^{g} H(\beta_i), \qquad (2.3)$$

as follows from the definition of the rational function $H(X)$. We then rewrite Equation (2.3) as follows

$$N_1 \cdot h(q^{-1/2}) = h(q^{-1/2}) + h(q^{1/2}) + g - R,$$

with

$$R = \sum_{i=1}^{g} H(\beta_i) + \sum_{r=1}^{m} (N_r - N_1) c_r q^{-r/2}.$$

Since $N_r \geq N_1$, $c_r \geq 0$ and $H(\beta_i) \geq 0$, it follows that $R \geq 0$ and hence

$$N_1 \cdot h(q^{-1/2}) \leq h(q^{-1/2}) + h(q^{1/2}) + g.$$

$\square$

Now we can prove:

Theorem 2.5. (Drinfeld-Vladut bound) (see [12]).   *The following bound holds:*

$$A(q) \leq \sqrt{q} - 1.$$

In particular we have for any tower $\mathcal{F}$ over $\mathbb{F}_q$:

$$\lambda(\mathcal{F}) \leq \sqrt{q} - 1.$$

Proof. For each $m \in \mathbb{N}$ with $m \geq 2$ we consider the polynomial $h_m(X) \in \mathbb{R}[X]$ which is given by

$$h_m(X) = \sum_{r=1}^{m} \left(1 - \frac{r}{m}\right) \cdot X^r. \tag{2.4}$$

The key point of the proof of Theorem 2.5 is the following equality

$$h_m(X) = \frac{X}{(X-1)^2} \cdot \left(\frac{X^m - 1}{m} + 1 - X\right), \tag{2.5}$$

which we prove in Lemma 2.6 below. For the associated rational function $H_m(X)$ we then get

$$
\begin{aligned}
H_m(X) &= 1 + h_m(X) + h_m(X^{-1}) \\
&= 1 + \frac{X}{(X-1)^2}\left(\frac{X^m - 1}{m} + 1 - X\right) \\
&\quad + \frac{X^{-1}}{(X^{-1} - 1)^2}\left(\frac{X^{-m} - 1}{m} + 1 - X^{-1}\right) \\
&= \frac{X}{(X-1)^2} \cdot \frac{X^m + X^{-m} - 2}{m} \\
&= \frac{2 - (X^m + X^{-m})}{m(X-1)(X^{-1} - 1)}.
\end{aligned}
\tag{2.6}
$$

For any complex number $\beta \neq 1$ with $|\beta| = 1$, the numbers $(\beta - 1)(\beta^{-1} - 1)$ and $2 - (\beta^m + \beta^{-m})$ are positive real numbers. Hence the hypothesis in Proposition 2.4 is satisfied; i.e., we have $H_m(\beta) \geq 0$ for all $\beta \in \mathbb{C}$ with

$|\beta| = 1$. It follows from Proposition 2.4 that for any function field F over $\mathbb{F}_q$ with genus $g(F) > 0$ the following inequality holds for all $m \geq 2$

$$\frac{N(F)}{g(F)} \leq \frac{1}{h_m(q^{-1/2})} + \frac{1}{g(F)} \cdot \left(1 + \frac{h_m(q^{1/2})}{h_m(q^{-1/2})}\right). \qquad (2.7)$$

Using again Equation (2.5) we see that

$$\lim_{m \to \infty} \frac{1}{h_m(q^{-1/2})} = \sqrt{q} - 1.$$

Let $\epsilon > 0$ be a real number and choose $n = n(\epsilon)$ such that

$$\frac{1}{h_n(q^{-1/2})} \leq \sqrt{q} - 1 + \epsilon/2.$$

Choose $g_0 = g_0(\epsilon, n)$ such that

$$\frac{1}{g_0} \cdot \left(1 + \frac{h_n(q^{1/2})}{h_n(q^{-1/2})}\right) < \epsilon/2.$$

Then we conclude from (2.7) with $m = n = n(\epsilon)$ that for all function fields $F/\mathbb{F}_q$ with $g(F) \geq g_0$,

$$\frac{N(F)}{g(F)} \leq (\sqrt{q} - 1 + \epsilon/2) + \epsilon/2 = \sqrt{q} - 1 + \epsilon.$$

$\square$

We still have to prove Equation (2.5):

Lemma 2.6. *For all $m \geq 2$ the following identity holds:*

$$\sum_{r=1}^{m} \left(1 - \frac{r}{m}\right) \cdot X^r = \frac{X}{(X-1)^2} \cdot \left(\frac{X^m - 1}{m} + 1 - X\right).$$

Proof. We set

$$f(X) := \sum_{r=1}^{m} X^r = \frac{X^{m+1} - X}{X - 1};$$

then we have

$$\frac{X \cdot f'(X)}{m} = \sum_{r=1}^{m} \frac{r}{m} \cdot X^r,$$

and therefore

$$\sum_{r=1}^{m} \left(1 - \frac{r}{m}\right) \cdot X^r = f(X) - \frac{X \cdot f'(X)}{m}$$

$$= \frac{X}{X-1} \cdot (X^m - 1) - \frac{X}{m} \cdot \frac{(X-1)((m+1)X^m - 1) - (X^{m+1} - X)}{(X-1)^2}$$

$$= \frac{X}{(X-1)^2} \cdot \left(\frac{X^m - 1}{m} + 1 - X\right).$$

$$\square$$

The interest in the quantity $A(q)$ also arose from applications of function fields to coding theory, cf. [48, 54]. The Tsfasman-Vladut-Zink theorem establishes a close connection between the asymptotics of $\mathbb{F}_q$-function fields (represented by the quantity $A(q)$) and the asymptotics of codes over $\mathbb{F}_q$. Some connections to cryptography are discussed in Chapter 2. For further connections to other areas we refer to [2, 40, 41, 44, 50, 52, 59].

Let us briefly recall the connection to coding theory. A linear code C over $\mathbb{F}_q$ of length $n = n(C)$ is a linear subspace of $\mathbb{F}_q^n$. The dimension $k = k(C)$ of C is its dimension as a vector space over $\mathbb{F}_q$. An important parameter of a linear code $C \neq \{0\}$ is its minimum distance $d = d(C)$, which is defined by

$$d = \min \{\text{wt}(c) \mid c \in C \text{ and } c \neq 0\},$$

where for a nonzero vector $c = (c_1, \ldots, c_n) \in \mathbb{F}_q^n$ its weight $\text{wt}(c)$ is given by

$$\text{wt}(c) = \#\{i \mid 1 \leq i \leq n \text{ and } c_i \neq 0\}.$$

A linear code C over $\mathbb{F}_q$ of length $n = n(C)$, dimension $k = k(C)$ and minimum distance $d = d(C)$ is briefly called an $[n, k, d]$-code, and the integers n, k and d are called the parameters of the code. In order to compare codes of different lengths, one also introduces relative parameters of the code C as follows:

- the *transmission rate* $R(C)$, given by $R(C) = k(C)/n(C)$.

- the *relative minimum distance* $\delta(C)$, given by $\delta(C) = d(C)/n(C)$.

We then get a map $\varphi : \{\mathbb{F}_q\text{-linear codes}\} \to [0, 1] \times [0, 1]$ by setting

$$C \xrightarrow{\varphi} (\delta(C), R(C)).$$

For a real number $\delta \in [0, 1]$ we consider the accumulation points of the image of the map φ on the vertical line $X = \delta$. The largest second coordinate of such accumulation points on the line $X = \delta$ is denoted by $\alpha_q(\delta)$. We can now state the connection of the asymptotics of codes (represented by $\alpha_q(\delta)$) with the quantity $A(q)$ that represents the asymptotics of $\mathbb{F}_q$-function fields:

Theorem 2.7. (Tsfasman-Vladut-Zink) (see [54], [48, p.207]). *Let q be a prime power such that $A(q) > 1$. Then*

$$\alpha_q(\delta) \geq 1 - A(q)^{-1} - \delta.$$

This result ensures the existence of arbitrary long codes (i.e., codes of arbitrary large length) having good parameters. For many values of q, Theorem 2.7 improves on the so-called Gilbert-Varshamov bound, which is a bound known from elementary coding theory, see [36].

Theorem 2.7 asks for good lower bounds for $A(q)$. For arbitrary q one knows that $A(q) > c \cdot \log q > 0$ with a real constant $c > 0$, see [47]. The actual value of $A(q)$ is only known when $q = \ell^2$ is a square. In this case we have the following result (see [31, 54] and Sections 4 and 5 below):

$$A(\ell^2) = \ell - 1, \quad \text{for any prime power } \ell.$$

This shows that the Drinfeld-Vladut bound given in Theorem 2.5 is sharp for finite fields of square cardinality. If $q = \ell^3$ is a cube, then we have the following good lower bound (see [8, 61] and also Section 4):

$$A(\ell^3) \geq \frac{2(\ell^2 - 1)}{\ell + 2}, \quad \text{for any prime power } \ell.$$

Much less is known about the quantity $A(\ell^r)$ for prime exponents $r \geq 5$.

Usually one gets information about the quantity $A(q)$ through the limits of towers of function fields over $\mathbb{F}_q$. The towers which appear in the literature are of the following three types:

- class field towers, cf. [43, 47];

- modular towers, cf. [14, 16, 31, 54];

- explicit towers, cf. [14, 21, 24].

By an explicit tower we mean a tower $\mathcal{F} = (F_0, F_1, F_2, \ldots)$ where each of the function fields F_i is given by explicit polynomial equations. For practical applications in coding theory and cryptography one needs an explicit description of the underlying function fields and of their $\mathbb{F}_q$-rational places.

Here we will mainly deal with explicit towers. Even more, the explicit description of the function fields $F_0, F_1, F_2, \ldots$ in the tower $\mathcal{F}$ will often have the following very simple shape:

Definition 2.8. Let $\mathcal{F} = (F_0, F_1, F_2, \ldots)$ be a tower of function fields over $\mathbb{F}_q$, where $F_0 = \mathbb{F}_q(x_0)$ is the rational function field. We say that the tower $\mathcal{F}$ is *recursive* if there exist a polynomial $f(X, Y) \in \mathbb{F}_q[X, Y]$ and functions $x_n \in F_n$ such that:

i) $f(X, Y)$ is separable in both variables X and Y;

ii) $F_{n+1} = F_n(x_{n+1})$ with $f(x_n, x_{n+1}) = 0$, for all $n \geq 0$;

iii) $[F_{n+1} : F_n] = \deg_Y f(X, Y)$, for all $n \geq 0$.

We also say that the tower $\mathcal{F}$ is given by the equation $f(X, Y) = 0$ or that $\mathcal{F}$ is defined recursively by the polynomial $f(X, Y)$. Sometimes a tower is recursively given by an equation of the form

$$g(X, Y) = h(X, Y) \tag{2.8}$$

with rational functions $g(X, Y)$ and $h(X, Y) \in \mathbb{F}_q(X, Y)$. It is obvious that, after clearing denominators, Equation (2.8) can be transformed into the form $f(X, Y) = 0$ with a polynomial $f(X, Y) \in \mathbb{F}_q[X, Y]$. For example, the defining equation

$$Y^\ell - Y = X^\ell / (X^{\ell-1} + 1)$$

can be transformed into the polynomial equation

$$f(X, Y) = (X^{\ell-1} + 1) \cdot (Y^\ell - Y) - X^\ell = 0,$$

cf. Section 5.1 below.

For a recursive tower $\mathcal{F} = (F_0, F_1, F_2, \ldots)$, much information about it is already contained in the field $F_1 = \mathbb{F}_q(x_0, x_1)$. So we define:

Definition 2.9. Let $\mathcal{F}$ be a recursive tower over $\mathbb{F}_q$ given by the polynomial equation $f(X, Y) = 0$. Then its *basic function field* is defined as

$$F = \mathbb{F}_q(x, y), \quad \text{with the relation } f(x, y) = 0.$$

It will be shown in Section 6.2 that for a recursive tower $\mathcal{F}$ with positive limit $\lambda(\mathcal{F}) > 0$ one has

$$\deg_X f(X, Y) = \deg_Y f(X, Y).$$

For the corresponding basic function field $F = \mathbb{F}_q(x, y)$ this condition means that

$$[F : \mathbb{F}_q(x)] = [F : \mathbb{F}_q(y)] \quad \text{if } \lambda(\mathcal{F}) > 0.$$

We recall the concepts of tame and wild ramification (see [48, p. 94]).

Definition 2.10. Let E/F be a function field extension. A place Q of the field E is *tamely ramified* (or *tame*) in the extension E/F, if the characteristic p does not divide the ramification index $e(Q|P)$, where P is the restriction of Q to the field F. Otherwise we say that Q is *wild* in the extension E/F. The extension E/F is called *tame* if all places of E are tame in E/F.

For example, a Galois extension E/F whose degree is relatively prime to the characteristic is a tame extension. This is the case for Kummer extensions (see Section 4). On the contrary, in the case of Artin-Schreier extensions (see Section 5) we have that all ramified places are wild.

The most convenient way to work with recursive towers is to think of them as pyramids; i.e., one considers in the same picture the fields

$$\mathbb{F}_q(x_n, x_{n+1}, \ldots, x_m), \quad \text{for all natural numbers } n \leq m.$$

We illustrate this way of thinking of a recursive tower with Figure 2.1 (see next page) that reaches the 8th step of the tower. The tower itself appears on the left edge of the pyramid.

For instance, the fields E and H in Figure 2.1 are $E = \mathbb{F}_q(x_1, x_2, x_3, x_4)$ and $H = \mathbb{F}_q(x_2, x_3, x_4, x_5, x_6)$. All fields on the same horizontal line are isomorphic to each other (for example, $F_3 \simeq E \simeq E'$ and $F_4 \simeq H$).

Let Q be a place of the field F_8, just for reasoning in the concrete situation of Figure 2.1. For the determination of the genus $g(F_8)$ one is led, by Hurwitz genus formula, to consider the ramification indices of (the restrictions of) Q in the various field extensions in Figure 2.1; i.e., in the extensions F_8/F_7, F_8/F_0, $G/E, G/H$, etc. One starts from the extensions at the base of the pyramid; i.e., from the extensions

$$\mathbb{F}_q(x_n, x_{n+1})/\mathbb{F}_q(x_n) \quad \text{and} \quad \mathbb{F}_q(x_n, x_{n+1})/\mathbb{F}_q(x_{n+1}),$$

with $0 \leq n \leq 7$. Knowing ramification indices in the extensions $F/\mathbb{F}_q(x)$ and $F/\mathbb{F}_q(y)$, where $F = \mathbb{F}_q(x, y)$ is the corresponding basic function field, one gets the ramification indices of the place Q at the base of the pyramid from the values $x_0(Q), x_1(Q), \ldots, x_8(Q)$. Then one tries to climb up the pyramid to the right and to the left by using repeatedly the following fundamental tool:

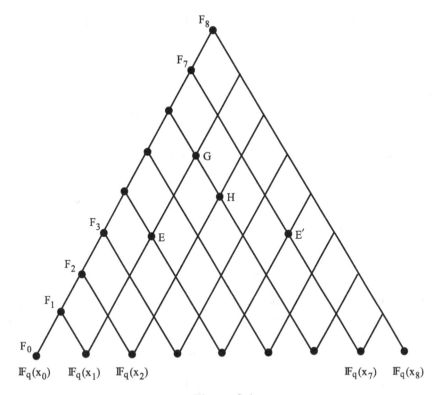

Figure 2.1

Theorem 2.11. (Abhyankar's lemma) (see [48, p.125]).   *Let E/F be a function field extension and let E_1, E_2 be two intermediate fields with $E = E_1 \cdot E_2$ (i.e., E is the composite field of E_1 and E_2). Let Q be a place of E and denote by Q_1, Q_2 and P its restrictions to E_1, E_2 and F. If $Q_1|P$ or $Q_2|P$ is tame, then*

$$e(Q|P) = \mathrm{lcm}(e(Q_1|P), e(Q_2|P)),$$

where lcm *stands for the least common multiple.*

Let us consider again the situation as in Figure 2.1. Suppose that all ramifications at the base of the pyramid are tame. It is then obvious that one gets easily all ramification indices in the pyramid by using Abhyankar's lemma repeatedly.

The situation is more difficult if wild ramification occurs at the base of the pyramid. This is in fact one of the major problems in dealing with the so-called wild towers (see Section 5): all known examples of explicit wild towers $\mathcal{F}$ with $\lambda(\mathcal{F}) > 0$ are such that the corresponding pyramids have infinitely many times the phenomenon illustrated in the following picture, where ℓ is a power of the characteristic and moreover $e(Q_1|P) = e(Q_2|P) = \ell$ (with notations as in Theorem 2.11).

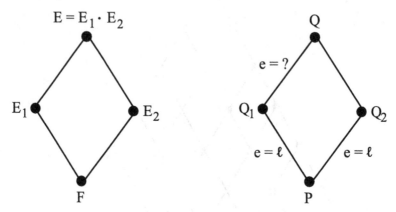

Figure 2.2

Abhyankar's lemma does not apply in this situation, and it is in general a hard task to determine the ramification index $e(Q|Q_1) = ?$. Even harder is in general the determination of the different exponent $d(Q|Q_1)$ of $Q|Q_1$. We will discuss some special cases of this situation in Section 5.

3.   Genus and Splitting Rate of a Tower

As before let $\mathcal{F} = (F_0, F_1, F_2, \ldots)$ be a tower of function fields over $\mathbb{F}_q$. We want to investigate the limit $\lambda(\mathcal{F}) = \lim_{n\to\infty} N(F_n)/g(F_n)$ of the tower (this limit does exist, see Proposition 3.2 and Definition 3.4). It will be convenient to treat the number of rational places and the genus separately.

Lemma 3.1. *Let $F_0 \subseteq F \subseteq E$ be finite separable extensions of algebraic function fields over $\mathbb{F}_q$. Then we have*

$$\frac{N(F)}{[F : F_0]} \geq \frac{N(E)}{[E : F_0]} \quad and \quad \frac{g(F) - 1}{[F : F_0]} \leq \frac{g(E) - 1}{[E : F_0]}.$$

In particular, if $g(F) \geq 2$ then

$$\frac{N(F)}{g(F) - 1} \geq \frac{N(E)}{g(E) - 1}.$$

Proof. It is clear that $N(E) \leq [E : F] \cdot N(F)$. Dividing this inequality by the degree $[E : F_0] = [E : F] \cdot [F : F_0]$ we get

$$\frac{N(E)}{[E : F_0]} \leq \frac{N(F)}{[F : F_0]}.$$

Since the extension E/F is separable, the Hurwitz genus formula gives that

$$2g(E) - 2 = [E : F] \cdot (2g(F) - 2) + \deg \mathrm{Diff}(E/F)$$

$$\geq [E : F] \cdot (2g(F) - 2).$$

We divide by $2[E : F_0]$ to obtain the desired result.   □

Proposition 3.2. *Given a tower* $\mathcal{F} = (F_0, F_1, F_2, \ldots)$ *over* $\mathbb{F}_q$, *the following limits do exist:*

$$\nu(\mathcal{F}) := \lim_{n\to\infty} N(F_n)/[F_n : F_0] \ \text{and} \ \gamma(\mathcal{F}) := \lim_{n\to\infty} g(F_n)/[F_n : F_0].$$

Proof. By Lemma 3.1, the sequence $(N(F_n)/[F_n : F_0])_{n\geq0}$ is monotonously decreasing, hence convergent in $\mathbb{R}$. On the other hand, we have that the sequence $((g(F_n) - 1)/[F_n : F_0])$ is monotonously increasing, hence convergent in $\mathbb{R} \cup \{\infty\}$. Since

$$\lim_{n\to\infty} (g(F_n) - 1)/[F_n : F_0] = \lim_{n\to\infty} g(F_n)/[F_n : F_0],$$

Proposition 3.2 follows.   □

Definition 3.3. The quantities $\nu(\mathcal{F})$ and $\gamma(\mathcal{F})$ in Proposition 3.2 are called the *splitting rate* and the *genus of the tower* $\mathcal{F}$, respectively.

One has that
$$0 \leq \nu(\mathcal{F}) \leq N(F_0) \ \text{and} \ 0 < \gamma(\mathcal{F}) \leq \infty.$$

If $\gamma(\mathcal{F}) < \infty$, we say that the tower has *finite genus*.

It follows from Proposition 3.2 that the sequence $N(F_n)/g(F_n)$ is convergent, since we have

$$\frac{N(F_n)}{g(F_n)} = \frac{N(F_n)/[F_n : F_0]}{g(F_n)/[F_n : F_0]} \to \frac{\nu(\mathcal{F})}{\gamma(\mathcal{F})} \ , \text{as } n \to \infty.$$

This leads us to the following definition which is crucial for the theory of towers.

Definition 3.4. For any tower $\mathcal{F} = (F_0, F_1, F_2, \ldots)$ over $\mathbb{F}_q$, the limit

$$\lambda(\mathcal{F}) := \lim_{n\to\infty} N(F_n)/g(F_n)$$

is called the *limit of the tower*.

We know from Section 2 that $0 \leq \lambda(\mathcal{F}) \leq A(q) \leq \sqrt{q} - 1$ (the last inequality is the Drinfeld-Vladut bound). Recall that the tower $\mathcal{F}$ is said to be asymptotically good if $\lambda(\mathcal{F}) > 0$; it is asymptotically optimal if $\lambda(\mathcal{F}) = A(q)$. We clearly have:

Corollary 3.5. *For a tower $\mathcal{F}$ over $\mathbb{F}_q$ one has that*

$$\lambda(\mathcal{F}) = \nu(\mathcal{F})/\gamma(\mathcal{F}).$$

Moreover, the following statements are equivalent:

i) The tower $\mathcal{F}$ is asymptotically good.

ii) The genus $\gamma(\mathcal{F})$ is finite and the splitting rate $\nu(\mathcal{F})$ is strictly positive.

Let $\mathcal{E} = (E_0, E_1, E_2, \ldots)$ and $\mathcal{F} = (F_0, F_1, F_2, \ldots)$ be two towers over $\mathbb{F}_q$. We call $\mathcal{E}$ a subtower of $\mathcal{F}$, if for any E_n there exists some F_m such that $E_n \subseteq F_m$.

Proposition 3.6. *If $\mathcal{E}$ is a subtower of $\mathcal{F}$, then $\lambda(\mathcal{E}) \geq \lambda(\mathcal{F})$. In particular, if the tower $\mathcal{F}$ is asymptotically good (resp. optimal), then any subtower $\mathcal{E}$ of $\mathcal{F}$ is also asymptotically good (resp. optimal).*

Proof. Let $E_n \subseteq F_m$, and suppose that $g(E_n) \geq 2$ (which holds for sufficiently large n, since $\mathcal{E}$ is a tower). By Lemma 3.1 we have

$$\frac{N(E_n)}{g(E_n) - 1} \geq \frac{N(F_m)}{g(F_m) - 1},$$

and hence $\lambda(\mathcal{E}) \geq \lambda(\mathcal{F})$.   □

In order to study the limit $\lambda(\mathcal{F})$ of a tower $\mathcal{F}$, it is often suitable to investigate separately the genus and the splitting rate of $\mathcal{F}$. We start with the investigation of the genus.

Definition 3.7. Let $\mathcal{F} = (F_0, F_1, F_2, \ldots)$ be a tower over $\mathbb{F}_q$, and let P be a place of F_0. We say that P is *ramified in the tower* $\mathcal{F}$ if for some $n \geq 1$ there exists a place Q of F_n lying above P such that $Q|P$ is ramified; i.e., the ramification index satisfies $e(Q|P) > 1$. If there exists an index $n \geq 1$ and a place Q of F_n above P such that $Q|P$ is wildly ramified (i.e., the characteristic of $\mathbb{F}_q$ divides the ramification index $e(Q|P)$), then P is said to be *wildly ramified in the tower* $\mathcal{F}$. Otherwise, the place P is said to be *tame* in $\mathcal{F}$. The set

$$V(\mathcal{F}) := \{P \mid P \text{ is a place of } F_0 \text{ which is ramified in } \mathcal{F}\}$$

is called the *ramification locus* of $\mathcal{F}$.

All asymptotically good towers which are known at present have a finite ramification locus. However, there are examples of non-recursive towers $\mathcal{F}$ over $\mathbb{F}_q$ such that the ramification locus $V(\mathcal{F})$ is infinite and the genus $\gamma(\mathcal{F})$

is finite, see [13]. A tower $\mathcal{F}$ with finite ramification locus $V(\mathcal{F})$ may have infinite genus $\gamma(\mathcal{F}) = \infty$, but in many cases one can use the next theorem to ensure the finiteness of $\gamma(\mathcal{F})$.

Recall the following notations: Let E/F be a finite separable extension of function fields, P a place of F and Q a place of E lying above P, then $e(Q|P)$ (resp. $d(Q|P)$) denotes the ramification index (resp. the different exponent) of the place Q over P.

Theorem 3.8. *Let $\mathcal{F} = (F_0, F_1, F_2, \ldots)$ be a tower with a finite ramification locus $V(\mathcal{F})$. Suppose that for each place $P \in V(\mathcal{F})$ there exists a real constant $c_P > 0$ such that, for all $n \geq 1$ and for all places Q of F_n lying above P, we have*

$$d(Q|P) \leq c_P \cdot e(Q|P).$$

Then the genus of the tower is finite and it satisfies

$$\gamma(\mathcal{F}) \leq g(F_0) - 1 + \frac{1}{2} \cdot \sum_{P \in V(\mathcal{F})} c_P \cdot \deg P < \infty.$$

Proof. The Hurwitz genus formula for the extension F_n/F_0 gives

$$2g(F_n) - 2 = (2g(F_0) - 2) \cdot [F_n : F_0] + \deg \mathrm{Diff}(F_n|F_0)$$

$$= (2g(F_0) - 2)[F_n : F_0] + \sum_{P \in V(\mathcal{F})} \sum_{Q|P} d(Q|P) \cdot \deg Q$$

$$\leq (2g(F_0) - 2) \cdot [F_n : F_0] + \sum_{P \in V(\mathcal{F})} c_P \cdot \sum_{Q|P} e(Q|P) \cdot \deg Q$$

$$= [F_n : F_0] \cdot \left(2g(F_0) - 2 + \sum_{P \in V(\mathcal{F})} c_P \cdot \deg P \right),$$

where the last equality follows from the "fundamental equality"

$$\sum_{Q|P} e(Q|P) \cdot f(Q|P) = [F_n : F_0],$$

see [48, p.64]. Dividing the inequality above by $2 \cdot [F_n : F_0]$ and letting $n \to \infty$, we obtain the desired result. $\qquad \square$

An important special case of Theorem 3.8 is the case of tame towers:

Corollary 3.9. *Let $\mathcal{F} = (F_0, F_1, F_2, \ldots)$ be a tower with a finite ramification locus $V(\mathcal{F})$, and suppose that all places $P \in V(\mathcal{F})$ are tame in $\mathcal{F}$. Then*

$$\gamma(\mathcal{F}) \leq g(F_0) - 1 + \frac{1}{2} \cdot \sum_{P \in V(\mathcal{F})} \deg P.$$

Proof. By Dedekind's different theorem, the different exponent of a tamely ramified place $Q|P$ satisfies $d(Q|P) = e(Q|P) - 1$, and hence we can choose $c_P := 1$ for each place $P \in V(\mathcal{F})$. ☐

In Section 5 we will see that Theorem 3.8 can also be applied to some interesting wild towers.

How can one check if the ramification locus $V(\mathcal{F})$ is finite? We discuss this problem in a particular case. Assume that the tower $\mathcal{F} = (F_0, F_1, F_2, \ldots)$ is recursively defined by the equation

$$\varphi(Y) = \psi(X), \qquad (3.1)$$

where $\varphi(T), \psi(T) \in \mathbb{F}_q(T)$ are rational functions (see Definition 2.8). As before, let $F = \mathbb{F}_q(x, y)$ be the corresponding basic function field which is given by the equation $\varphi(y) = \psi(x)$, and define

$$V_0 := \{P \mid P \text{ is a place of } \mathbb{F}_q(x) \text{ which ramifies in } F/\mathbb{F}_q(x)\}.$$

The set V_0 is finite, since the extension $F/\mathbb{F}_q(x)$ is separable. We also define

$$R_0 := \{x(P) \mid P \in V_0\}. \qquad (3.2)$$

Clearly, this set R_0 is a finite subset of $\overline{\mathbb{F}}_q \cup \{\infty\}$.

Proposition 3.10. *Let* $\mathcal{F} = (F_0, F_1, F_2, \ldots)$ *be a tower over* $\mathbb{F}_q$ *which is recursively defined by Equation* (3.1). *Assume that there exists a finite subset* $R \subseteq \overline{\mathbb{F}}_q \cup \{\infty\}$ *such that the following two conditions hold:*

a) The set R contains R_0, with R_0 as in Equation (3.2).

b) If $\beta \in R$ and $\alpha \in \overline{\mathbb{F}}_q \cup \{\infty\}$ satisfy the equation $\varphi(\beta) = \psi(\alpha)$, then $\alpha \in R$.

Then the ramification locus of the tower $\mathcal{F}$ satisfies

$$V(\mathcal{F}) \subseteq \{P \mid P \text{ is a place of } F_0 \text{ with } x_0(P) \in R\};$$

in particular, $V(\mathcal{F})$ is finite and moreover

$$\sum_{P \in V(\mathcal{F})} \deg P \leq \#R. \qquad (3.3)$$

Proof. Let $P \in V(\mathcal{F})$. There is some $n \geq 0$ and a place Q of F_n lying above P such that Q is ramified in the extension F_{n+1}/F_n. Let $P' := Q \cap \mathbb{F}_q(x_n)$ denote the place of $\mathbb{F}_q(x_n)$ lying below Q, and consider the following diagram:

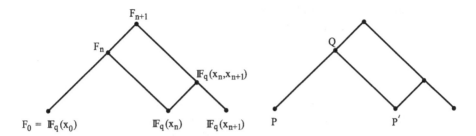

Figure 3.1

Since Q is ramified in F_{n+1}/F_n, the place P' is ramified in the extension $\mathbb{F}_q(x_n, x_{n+1})$ over $\mathbb{F}_q(x_n)$. Hence $\beta := x_n(P') \in R_0$. For $i = 0, \ldots, n$ we set $\alpha_i := x_i(Q)$; then (by Condition a)) we have that $\alpha_n = \beta \in R$. Since

$$\varphi(\alpha_i) = \psi(\alpha_{i-1}) \text{ for each } i = 1, \ldots, n,$$

it follows from Condition b) that $\alpha_{n-1}, \alpha_{n-2}, \ldots, \alpha_0 \in R$, and in particular we have $x_0(P) = x_0(Q) = \alpha_0 \in R$. We have thus shown that $P \in V(\mathcal{F})$ implies $x_0(P) \in R$. In order to prove Inequality (3.3) one just notes that $\sum_{P \in V(\mathcal{F})} \deg P$ is invariant under constant field extensions. $\qquad \square$

Now we start the investigation of the splitting rate. In particular we want to establish a criterion analogous to Proposition 3.10 which implies a positive splitting rate $\nu(\mathcal{F}) > 0$, for a particular class of recursive towers.

Definition 3.11. Let $\mathcal{F} = (F_0, F_1, F_2, \ldots)$ be a tower over $\mathbb{F}_q$, and let P be a rational place of F_0 (i.e., $\deg P = 1$). We say that P *splits completely in the tower* $\mathcal{F}$ if P splits completely in all extensions F_n/F_0; i.e., there are exactly $[F_n : F_0]$ places of F_n above the place P (and they are rational places of F_n). The set

$$Z(\mathcal{F}) := \{P \mid P \text{ is a rational place of } F_0 \text{ which splits completely in } \mathcal{F}\}$$

is called the *splitting locus of the tower* $\mathcal{F}$. It is obvious that $Z(\mathcal{F}) \cap V(\mathcal{F}) = \emptyset$.

Lemma 3.12. *Let $\mathcal{F}$ be a tower over $\mathbb{F}_q$. Then we have*

$$\nu(\mathcal{F}) \geq \#Z(\mathcal{F}).$$

Proof. The assertion is trivial since for all n, any place $P \in Z(\mathcal{F})$ has $[F_n : F_0]$ distinct extensions in the field F_n, all of them being rational places of F_n. $\quad \square$

For a rational function field $\mathbb{F}_q(z)$ and an element $\alpha \in \mathbb{F}_q$, we denote by $(z = \alpha)$ the place which is the zero of the function $z - \alpha$ in $\mathbb{F}_q(z)$. Similarly,

$(z = \infty)$ denotes the pole of the function z in $\mathbb{F}_q(z)$. We consider again a tower $\mathcal{F}$ over $\mathbb{F}_q$ which is defined recursively by the equation

$$\varphi(Y) = \psi(X). \tag{3.4}$$

Proposition 3.13. *Let $\mathcal{F} = (F_0, F_1, F_2, \ldots)$ be a tower over $\mathbb{F}_q$ defined recursively by Equation (3.4), and let $F = \mathbb{F}_q(x, y)$ be the corresponding basic function field with the relation $\varphi(y) = \psi(x)$. Assume that there exists a nonempty subset S of $\mathbb{F}_q \cup \{\infty\}$ which satisfies the following two conditions:*

a) *For all $\alpha \in S$, the place $(x = \alpha)$ of $\mathbb{F}_q(x)$ splits completely in the extension $F/\mathbb{F}_q(x)$.*

b) *If $\alpha \in S$ and if Q is a place of F lying above the place $(x = \alpha)$, then $y(Q) \in S$.*

Then for all $\alpha \in S$, the place $(x_0 = \alpha)$ of $F_0 = \mathbb{F}_q(x_0)$ splits completely in the tower $\mathcal{F}$; i.e., we have $(x_0 = \alpha) \in Z(\mathcal{F})$. In particular, the splitting rate satisfies

$$\nu(\mathcal{F}) \geq \#S.$$

Proof. By induction: Let $\alpha \in S$ and let Q be a place of F_n lying above the place $(x_0 = \alpha)$. Then $x_n(Q) =: \beta \in S$, by Condition b). The place $(x_n = \beta)$ splits completely in the extension $\mathbb{F}_q(x_n, x_{n+1})/\mathbb{F}_q(x_n)$, by Condition a). Therefore the place Q splits completely in the extension F_{n+1}/F_n. The inequality for the splitting rate follows from Lemma 3.12. $\square$

Note that both conditions a) and b) in Proposition 3.13 follow from the stronger condition below:

Condition c)   For all $\alpha \in S$, the equation $\varphi(T) = \psi(\alpha)$ has $m = \deg\varphi$ distinct roots in the set S.

For an absolutely irreducible polynomial $f(X, Y) \in \mathbb{F}_q[X, Y]$, it is in general not true that the equation $f(X, Y) = 0$ defines a recursive tower $\mathcal{F} = (F_0, F_1, F_2, \ldots)$; i.e., $F_0 = \mathbb{F}_q(x_0)$ is a rational function field and $F_{n+1} = F_n(x_{n+1})$ with the relation $f(x_n, x_{n+1}) = 0$. It may happen, for instance, that the fields defined in this way satisfy $F_r = F_{r+1} = F_{r+2} = \ldots$, for some index $r \geq 1$. However in many cases, the following proposition shows that the equation $f(X, Y) = 0$ defines indeed a recursive tower.

Proposition 3.14. *Let $f(X, Y) \in \mathbb{F}_q[X, Y]$ be a polynomial which satisfies the condition $\deg_Y f(X, Y) = m \geq 2$, and let $F_0 \subseteq F_1 \subseteq F_2 \subseteq \ldots$ be a sequence of function fields over $\mathbb{F}_q$, recursively defined by $F_0 = \mathbb{F}_q(x_0)$ and $F_{n+1} = F_n(x_{n+1})$ with $f(x_n, x_{n+1}) = 0$. Suppose that for each $n \geq 0$ there exists a place Q_n of F_n such that the following two conditions hold (see Figure 3.2 below):*

a) *There is a place R_n of the function field $\mathbb{F}_q(x_n, x_{n+1})$ lying above the place $P_n := Q_n \cap \mathbb{F}_q(x_n)$, such that the ramification index of $R_n|P_n$ satisfies $e(R_n|P_n) = m$.*

b) *The ramification index $e(Q_n|P_n)$ is relatively prime to m.*

Then $[F_{n+1} : F_n] = m$ for all $n \geq 0$, and the equation $f(X,Y) = 0$ defines a recursive tower $\mathcal{F}$ over $\mathbb{F}_q$.

Proof. We proceed by induction: Suppose that the field $\mathbb{F}_q$ is algebraically closed in F_n and consider the field extension F_{n+1}/F_n, with $F_{n+1} = F_n(x_{n+1})$ and $f(x_n, x_{n+1}) = 0$. We have the following situation, where P_n is the restriction of the place Q_n to $\mathbb{F}_q(x_n)$ and R_n is the unique place of $\mathbb{F}_q(x_n, x_{n+1})$ above P_n:

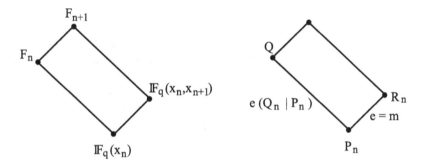

Figure 3.2

It follows from Abhyankar's lemma, that the place Q_n is ramified in F_{n+1}/F_n with ramification index m. In particular we have that $[F_{n+1} : F_n] \geq m$. On the other hand, the element x_{n+1} satisfies the equation $f(x_n, x_{n+1}) = 0$ over F_n and therefore $[F_{n+1} : F_n] = [F_n(x_{n+1}) : F_n] \leq m$. Hence $[F_{n+1} : F_n] = m$, the place Q_n is totally ramified in F_{n+1}/F_n and $\mathbb{F}_q$ is also algebraically closed in F_{n+1}. □

Remark 3.15. In many interesting cases (see Section 4 and Section 5), the polynomial $f(X,Y)$ guarantees a very special ramification behaviour at the base of the pyramid, which implies immediately that the equation $f(X,Y) = 0$ indeed defines a recursive tower. As before we set $m := \deg_Y f(X,Y) \geq 2$, and we assume that there exists a place P_0 of $\mathbb{F}_q(x_0) = F_0$ which leads to the ramification picture in Figure 3.3.

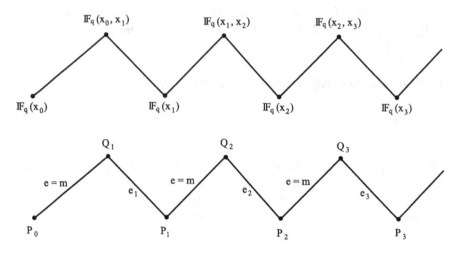

Figure 3.3

This picture means: the place P_0 of $\mathbb{F}_q(x_0)$ is ramified in the extension $\mathbb{F}_q(x_0, x_1)$ over $\mathbb{F}_q(x_0)$ with ramification index $e = m$. Hence there is just one place Q_1 of $\mathbb{F}_q(x_0, x_1)$ lying above P_0, and this place Q_1 is totally ramified over P_0. The place $P_1 := Q_1 \cap \mathbb{F}_q(x_1)$ of $\mathbb{F}_q(x_1)$ also has ramification index $e = m$ in the extension $\mathbb{F}_q(x_1, x_2)/\mathbb{F}_q(x_1)$, and we denote by Q_2 the unique place of $\mathbb{F}_q(x_1, x_2)$ lying above P_1, etc. Moreover we make the crucial assumption that the ramification indices $e_1 := e(Q_1|P_1)$, $e_2 := e(Q_2|P_2)$, $e_3 := e(Q_3|P_3)$, ... are all relatively prime to m. Using Abhyankar's lemma repeatedly as in the proof of Proposition 3.14 one concludes that $[F_n : F_0] = m^n$, that P_0 is totally ramified in F_n/F_0 and that the equation $f(X, Y) = 0$ indeed defines a recursive tower over $\mathbb{F}_q$.

Remark 3.16. There are recursive towers which do not satisfy the assumptions of Proposition 3.14. In such a case it seems to be more difficult to decide if the corresponding equation $f(X, Y) = 0$ defines indeed a tower (see [57, 58]).

4.   Explicit Tame Towers

Before presenting some explicit asymptotically good tame towers of function fields, we make the following general remark: Let $\mathcal{F} = (F_0, F_1, F_2, \ldots)$ be a recursive $\mathbb{F}_q$-tower, given by a polynomial $f(X, Y) \in \mathbb{F}_q[X, Y]$. Let $h(Z) \in \mathbb{F}_q(Z)$ be a fractional linear transformation; i.e., $h(Z) = (aZ + b)/(cZ + d)$ with $a, b, c, d \in \mathbb{F}_q$ and $ad \neq bc$. Then the tower $\mathcal{F}$ can also be defined by the equation

$$g(X, Y) := f(h(X), h(Y)) = 0.$$

Performing such a fractional linear transformation can sometimes transform the defining equation to a nicer form, or it can make it easier to describe the ramification locus or the splitting locus of the tower.

All towers T that we consider in this section are recursive tame towers, which satisfy the hypothesis of Proposition 3.10 and hence they have a finite ramification locus. Moreover they have a non-empty splitting locus $Z(T)$. Then we get from Section 3 the following lower bound for the limit $\lambda(T)$ of the tower T:

Lemma 4.1. *Assume that T is a recursive tame tower defined by Equation (3.4) and satisfying the hypothesis of Proposition 3.10. Then its limit $\lambda(T)$ satisfies the estimate*

$$\lambda(T) \geq \frac{2 \cdot \#Z(T)}{\#R - 2},$$

where R is the finite set whose existence is the main assumption made in Proposition 3.10.

Proof. We have the inequalities $\nu(T) \geq \#Z(T)$ (see Lemma 3.12) and also $\gamma(T) \leq (-2 + \#R)/2$ (see Corollary 3.9 and Proposition 3.10). The assertion follows then immediately since $\lambda(T) = \nu(T)/\gamma(T)$ (see Corollary 3.5). $\square$

The defining equations that we will consider in this section do give rise to towers of function fields $T = (F_0, F_1, F_2, \ldots)$, since it will always be the case that in all extensions F_{n+1}/F_n arising from our equations, there exist places of the field F_n that are totally ramified in F_{n+1} (see Proposition 3.14 and Remark 3.15).

4.1   The Tower T_1

Consider the tower T_1 over the field $\mathbb{F}_4$ with four elements, which is given recursively by the equation

$$Y^3 = X^3/(X^2 + X + 1). \tag{4.1}$$

Let $P = (x_0 = \infty)$ denote the place at infinity of the rational function field $F_0 = \mathbb{F}_4(x_0)$ and let Q denote a place of the field $F_1 = F_0(x_1)$ above P. The place P is a simple pole of the right hand side of the defining equation $x_1^3 = x_0^3/(x_0^2 + x_0 + 1)$, and we get

$$v_Q(x_1^3) = -e(Q|P); \quad \text{hence} \quad e(Q|P) = 3 \text{ and } v_Q(x_1) = -1.$$

Here v_Q denotes the valuation corresponding to the place Q, and $e(Q|P)$ is the ramification index of $Q|P$. This shows that the place P is totally ramified in the field F_1, and in particular that Q is the unique place of F_1 above P. Also, since $v_Q(x_1) = -1$, the place Q is a simple pole for the right hand side of the defining equation $x_2^3 = x_1^3/(x_1^2 + x_1 + 1)$, and we conclude as before that the place Q is totally ramified in the field extension F_2/F_1, and so on. In this way we see that the defining equation (4.1) really leads to a tower, since the place P is totally ramified in all extensions.

Now we show that the place $P_0 = (x_0 = 0)$ of F_0 splits completely in the tower. Let Q_0 be a place of F_1 above P_0. From the defining equation $x_1^3 = x_0^3/(x_0^2 + x_0 + 1)$, we see that $x_1(Q_0) = 0$. We have that

$$F_1 = F_0(x_1/x_0), \quad \text{with } (x_1/x_0)^3 = 1/(x_0^2 + x_0 + 1).$$

Since $x_0(P_0) = 0$, it follows from the last equation above that P_0 splits completely in the extension F_1/F_0. Again we have that

$$F_2 = F_1(x_2/x_1), \quad \text{with } (x_2/x_1)^3 = 1/(x_1^2 + x_1 + 1).$$

Since $x_1(Q_0) = 0$, it follows from the last equation above that each of the three places Q_0 of F_1 above P_0 splits completely in the extension F_2/F_1, and so on. Thus the splitting locus of the tower $\mathcal{T}_1$ has cardinality $\#Z(\mathcal{T}_1) \geq 1$.

>From the theory of Kummer extensions (see [48, p.110 ff.]), or with arguments similar to the ones used above for the place P, one obtains that the set R_0 in Proposition 3.10 (see Equation (3.2)) is given by $R_0 = (\mathbb{F}_4 \setminus \mathbb{F}_2) \cup \{\infty\}$; i.e., the elements of R_0 are the poles of the function $X^3/(X^2 + X + 1)$, and they are simple poles. We show now that the set $R = \mathbb{F}_4^\times \cup \{\infty\}$ satisfies Condition b) in Proposition 3.10. In fact if $\beta = \infty$ and $\alpha^3/(\alpha^2 + \alpha + 1) = \beta^3$, then $\alpha = \infty$ or $\alpha^2 + \alpha + 1 = 0$, hence $\alpha \in R_0$. If $\beta \in \mathbb{F}_4^\times$ then

$$\alpha^3/(\alpha^2 + \alpha + 1) = \beta^3 = 1,$$

and hence $\alpha^3 = \alpha^2 + \alpha + 1$. Since the characteristic is $p = 2$, we get $(\alpha + 1)^3 = \alpha^3 + \alpha^2 + \alpha + 1 = 0$, therefore $\alpha = 1 \in R$. From Lemma 4.1 we conclude now that the limit $\lambda(\mathcal{T}_1)$ satisfies

$$\lambda(\mathcal{T}_1) \geq \frac{2 \cdot 1}{4 - 2} = 1 = \sqrt{4} - 1.$$

So the tower $\mathcal{T}_1$ over the field $\mathbb{F}_4$ attains the Drinfeld-Vladut bound; i.e., it is asymptotically optimal over $\mathbb{F}_4$. $\square$

Performing the fractional linear transformation $h(Z) = 1/Z$, we see that the tower T_1 can also be defined by the nicer equation $Y^3 = (X + 1)^3 - 1$. The tower T_1 is therefore the very particular case $\ell = r = 2$ of the following tower T_2.

4.2 The Tower T_2

Let ℓ be any prime power and let $q = \ell^r$, where $r \in \mathbb{N}$ and $r \geq 2$. Consider the tower T_2 over $\mathbb{F}_q$ which is given recursively by the equation

$$Y^m = (X + 1)^m - 1, \quad \text{with } m = (q - 1)/(\ell - 1). \tag{4.2}$$

Similarly as for the tower T_1 one shows that the place $P_0 = (x_0 = 0)$ of the function field $F_0 = \mathbb{F}_q(x_0)$ is totally ramified in all extensions F_n/F_0, so that Equation (4.2) does define a recursive tower $T_2 = (F_0, F_1, F_2, \ldots)$. One can also argue as in Remark 3.15 as follows: In the ramification picture of Figure 3.3 for the place $P_0 = (x_0 = 0)$ one has that P_i is the zero of the function x_i in $\mathbb{F}_q(x_i)$ for all $i \geq 0$, and the ramification indices e_i in Figure 3.3 are all equal to $e_i = 1$ (as follows from the equation $x_i^m = (x_{i-1} + 1)^m - 1$). Therefore Equation (4.2) does define a tower, and the place P_0 is totally ramified in all extensions F_n/F_0.

Let $F = \mathbb{F}_q(x, y)$ with $y^m = (x + 1)^m - 1$ be the basic function field corresponding to the tower T_2 and let V_0 be the set of places of $\mathbb{F}_q(x)$ which ramify in $F/\mathbb{F}_q(x)$. The set $R_0 = \{x(P) \mid P \in V_0\}$ (as defined in (3.2)) is here given by

$$R_0 = \{\beta \in \overline{\mathbb{F}}_q \mid (\beta + 1)^m = 1\},$$

as follows from the theory of Kummer extensions of function fields. We claim that the set $R := \mathbb{F}_q$ satisfies the conditions in Proposition 3.10. In fact, we have $R_0 \subseteq \mathbb{F}_q$ (since $m = (q-1)/(\ell-1)$ is the norm exponent of the extension $\mathbb{F}_q/\mathbb{F}_\ell$), and for $\beta \in \mathbb{F}_q$ and $\alpha \in \overline{\mathbb{F}}_q$ with $\beta^m = (\alpha + 1)^m - 1$ it follows that $(\alpha + 1)^m = 1 + \beta^m \in \mathbb{F}_\ell$, hence $\alpha \in \mathbb{F}_q$. By Proposition 3.10, the ramification locus $V(T_2)$ is finite and it satisfies

$$V(T_2) \subseteq \{P \mid P \text{ is a place of } F_0 \text{ with } x_0(P) \in \mathbb{F}_q\}.$$

Next we show that the place $P_\infty = (x_0 = \infty)$ of the rational function field $F_0 = \mathbb{F}_q(x_0)$ splits completely in the tower T_2. We have

$$F_1 = F_0 \left(\frac{x_1}{x_0 + 1} \right) \quad \text{and} \quad \left(\frac{x_1}{x_0 + 1} \right)^m = 1 - \left(\frac{1}{x_0 + 1} \right)^m.$$

The right hand side of the last equality above takes the value 1 at the place P_∞, and since the exponent $m = (q - 1)/(\ell - 1)$ is the norm exponent of $\mathbb{F}_q/\mathbb{F}_\ell$

we conclude that P_∞ splits completely in the extension F_1/F_0. Let Q_∞ be a place of F_1 above P_∞. Then we have from the equation $x_1^m = (x_0 + 1)^m - 1$ that $v_{Q_\infty}(x_1) = v_{Q_\infty}(x_0) = -1$. Since

$$F_2 = F_1\left(\frac{x_2}{x_1 + 1}\right) \quad \text{and} \quad \left(\frac{x_2}{x_1 + 1}\right)^m = 1 - \left(\frac{1}{x_1 + 1}\right)^m,$$

it follows as above that the place Q_∞ splits completely in the extension F_2/F_1. Repeating this argument we find that P_∞ splits completely in the tower $\mathcal{T}_2$, thus $\#Z(\mathcal{T}_2) \geq 1$. From Lemma 3.1, we then get a positive limit

$$\lambda(\mathcal{T}_2) \geq 2/(q - 2) > 0;$$

i.e., the tower $\mathcal{T}_2$ over $\mathbb{F}_q$ is asymptotically good.   □

Using class field theory, J.-P. Serre [49] proved in particular that $A(q) > 0$ for all prime powers q. The tower $\mathcal{T}_2$ above gives a very simple proof of this result of Serre, when q is not a prime number. No asymptotically good explicit tower over a finite field of prime order is known at present, and it is one of the main challenges to construct an explicit asymptotically good tower over a prime field.

The tower $\mathcal{T}_2$ above is a special case of the so-called towers of Fermat type; these are defined as follows: Let $a, b, c \in \mathbb{F}_q^\times$ and let $m \geq 2$ be a natural number which is not divisible by $p = \text{char } \mathbb{F}_q$. Then the equation

$$Y^m = a(X + b)^m + c$$

does define a tower over $\mathbb{F}_q$, see [58]. Some of these towers of Fermat type can be shown to be asymptotically good, with similar arguments as in Section 4.2 above (see [24, 28]). For example, let ℓ be any prime power with $\ell \geq 3$ and let $q = \ell^2$. Take $m = \ell - 1$, choose any $b \in \mathbb{F}_\ell^\times$ and consider the tower $\mathcal{T}$ over $\mathbb{F}_q$ which is given recursively by the equation

$$Y^{\ell-1} = -(X + b)^{\ell-1} + 1.$$

Its limit satisfies $\lambda(\mathcal{T}) \geq 2/(\ell - 2)$. In particular for $\ell = 3$, we obtain a tower over the field $\mathbb{F}_9$ attaining the Drinfeld-Vladut bound. Using the transformation $h(Z) = b \cdot Z$, we see that all these towers (for distinct values of $b \in \mathbb{F}_\ell^\times$) are equal to each other.

4.3   The Tower $\mathcal{T}_3$

In this subsection we discuss another interesting tame tower that was introduced in [24]. Let p be an odd prime number and let $q = p^2$. Consider the equation

$$Y^2 = \frac{X^2 + 1}{2X} \tag{4.3}$$

over the field $\mathbb{F}_q$. Similarly to the case of the tower $\mathcal{T}_1$ in Section 4.1, one sees that Equation 4.3 does define recursively a tower $\mathcal{T}_3 = (F_0, F_1, F_2, \ldots)$ over $\mathbb{F}_q$, and that the place $P_\infty = (x_0 = \infty)$ of the rational function field $F_0 = \mathbb{F}_q(x_0)$ is totally ramified in all extensions F_n/F_0.

Let $a \in \mathbb{F}_q$ be such that $a^2 = -1$ (note that such an element exists in $\mathbb{F}_q$ since $q = p^2$). The set R_0 in Proposition 3.10 is here given by $R_0 = \{0, \infty, \pm a\}$. We claim that the set

$$R = \{0, \infty, \pm a, \pm 1\}$$

satisfies Condition b) in Proposition 3.10. This follows from the following assertions, which are all easily verified:

- if $\beta = 0$ and $\beta^2 = (\alpha^2 + 1)/2\alpha$, then $\alpha = \pm a$.

- if $\beta = \infty$ and $\beta^2 = (\alpha^2 + 1)/2\alpha$, then $\alpha \in \{0, \infty\}$.

- if $\beta = \pm a$ and $\beta^2 = (\alpha^2 + 1)/2\alpha$, then $\alpha = -1$.

- if $\beta = \pm 1$ and $\beta^2 = (\alpha^2 + 1)/2\alpha$, then $\alpha = 1$.

Much harder is here the determination of a single rational place of F_0 that splits completely over $\mathbb{F}_{p^2}$ in this tower $\mathcal{T}_3$. In fact, there are at least $2(p-1)$ such places; i.e., $\#Z(\mathcal{T}_3) \geq 2(p-1)$. From this and from Lemma 4.1 we then get

$$\lambda(\mathcal{T}_3) \geq \frac{2 \cdot 2(p-1)}{6 - 2} = p - 1;$$

i.e., the tower $\mathcal{T}_3$ over $\mathbb{F}_{p^2}$ attains the Drinfeld-Vladut bound. In particular, one concludes that $\#Z(\mathcal{T}_3) = 2(p-1)$.

For the description of the completely splitting places in the tower $\mathcal{T}_3$ the following polynomial $H(X) \in \mathbb{F}_p[X]$ plays a crucial role:

$$H(X) = \sum_{m=0}^{(p-1)/2} \binom{(p-1)/2}{m}^2 \cdot X^m,$$

where $\binom{(p-1)/2}{m}$ denotes the binomial coefficient. The polynomial $H(X)$ is the so-called Deuring polynomial; its roots parametrize supersingular elliptic curves in Legendre normal form (see [5]). It is well-known (but non-trivial) that $H(X)$ is a separable polynomial having all roots in the field $\mathbb{F}_{p^2}$. The key point here is to prove the following polynomial identity (see [24]):

$$H(X^4) = X^{p-1} \cdot H\left(\left(\frac{X^2+1}{2X}\right)^2\right). \tag{4.4}$$

A nice remark by M. Zieve is that it follows from Equation (4.4) that the roots of $H(X)$ are in fact 4-th powers in the field $\mathbb{F}_{p^2}$; i.e., we have the following inclusion

$$S := \{\alpha \in \overline{\mathbb{F}}_p | H(\alpha^4) = 0\} \subseteq \mathbb{F}_{p^2}. \tag{4.5}$$

Zieve's argument is as follows: If $H(\alpha^4) = 0$ then $H(((\alpha^2+1)/2\alpha)^2) = 0$ by Equation (4.4). Since all roots of $H(X)$ are in $\mathbb{F}_{p^2}$, it follows that $\alpha^4 \in \mathbb{F}_{p^2}$ and that $((\alpha^2+1)/2\alpha)^2 \in \mathbb{F}_{p^2}$, and then $\alpha^2 \in \mathbb{F}_{p^2}$. We have thus shown

$$H(\beta^2) = 0 \Rightarrow \beta \in \mathbb{F}_{p^2}.$$

In particular, since $H(((\alpha^2+1)/2\alpha)^2) = 0$, we obtain that $(\alpha^2+1)/2\alpha \in \mathbb{F}_{p^2}$. Since also $\alpha^2 \in \mathbb{F}_{p^2}$ we see that the element α itself is in $\mathbb{F}_{p^2}$. This proves that the set S in (4.5) is contained in $\mathbb{F}_{p^2}$ (for another proof see H. G. Rück's appendix to [24]). The cardinality of S is $\#S = 2(p-1)$, since $H(0) \neq 0$ and $H(X)$ is a separable polynomial.

It is now a simple matter to check (using Equation (4.4)) that the set S in (4.5) above satisfies Condition c) just after Proposition 3.13, and hence it follows from Proposition 3.13 that $\#Z(\mathcal{T}_3) \geq \#S = 2(p-1)$. $\square$

It follows from the work of N. Elkies [14] that the tower $\mathcal{T}_3$ is in fact the modular tower $X_0(2^n)$, see also [24, p.75 ff.].

The key identity Equation (4.4) satisfied by Deuring's polynomial is proved by using Gauss' hypergeometric differential equation. This idea of using certain differential equations to control rational places in tame towers was taken again by Beelen-Bouw, providing a more systematic technique for the search for asymptotically good tame towers. We just illustrate an application of their technique: If p is a prime number and $p \equiv \pm 1 \mod 8$, then the tower $\mathcal{T}$ over $\mathbb{F}_{p^2}$ which is defined recursively by the equation

$$Y^2 = \frac{X(1-X)}{X+1}$$

attains the Drinfeld-Vladut bound, see Proposition 4.6 in [3] and Example 4.5 in [24].

All tame towers considered here so far have "separated variables"; i.e., they are all defined recursively by an equation $\varphi(Y) = \psi(X)$ with two rational functions $\varphi(Y) \in \mathbb{F}_q(Y)$ and $\psi(X) \in \mathbb{F}_q(X)$. There are also very interesting tame towers with non-separated variables (see [33, 37]). For example, the four recursive towers over $\mathbb{F}_{p^2}$ (for $p = 2, 3, 5$ and 7) given by the following polynomials $f(X, Y)$ are asymptotically optimal:

- Case $p = 2$ and $f(X, Y) = X^2Y^3 + (X^3 + X^2 + X)Y^2 + (X + 1)Y + X^3 + X$.

- Case $p = 3$ and $f(X, Y) = 2XY^2 + (X^2 + X + 1)Y + X^2 + X + 2$.

- Case $p = 5$ and $f(X, Y) = (4X + 1)Y^2 + (X^2 + X + 2)Y + X + 3$.

- Case $p = 7$ and $f(X, Y) = (X^2 + 6)Y^2 + XY + X^2 + 4$.

All four towers above were shown to be elliptic modular by Elkies (see the appendix in [33]).

We finish this section with the remark that all optimal recursive towers presented here support the "modularity conjecture" for such towers; a conjecture which was proposed by N. Elkies in [14].

5. Explicit Wild Towers

In this section we discuss wild towers $\mathcal{F}$ over $\mathbb{F}_q$; i.e., towers such that there exist places which are wildly ramified in $\mathcal{F}$. As was pointed out at the end of Section 2, the calculation of the genus $\gamma(\mathcal{F})$ for wild towers is more difficult than in the case of tame towers, since Abhyankar's lemma does not apply in general (see Figure 2.2). Typical examples of wild towers are the so-called Artin-Schreier towers $\mathcal{F} = (F_0, F_1, F_2, \ldots)$. Here F_0 is the rational function field over $\mathbb{F}_q$, and all extensions F_{n+1}/F_n are Galois of degree $[F_{n+1} : F_n] = p = \mathrm{char}(\mathbb{F}_q)$. As follows from Galois theory, there then exist elements $x_{n+1} \in F_{n+1}$ and $z_n \in F_n$ such that $F_{n+1} = F_n(x_{n+1})$ and $x_{n+1}^p - x_{n+1} = z_n$, for all $n \geq 0$.

In some cases of wild ramification, Lemma 5.1 below replaces Abhyankar's lemma. It will be very useful in order to estimate the genus $\gamma(\mathcal{F})$ of some specific wild towers (see Theorem 5.7 and Theorem 5.17 below). First we recall some notations and facts: Let E/F be a finite separable extension of function fields, let P be a place of F and Q a place of E lying above P. Then $e(Q|P)$ and $d(Q|P)$ denote the ramification index and the different exponent of $Q|P$, respectively. If the extensions E/F is Galois of degree $p = \mathrm{char}(\mathbb{F}_q)$, it is well-known that $d(Q|P) = s \cdot (e(Q|P) - 1)$ for some $s \geq 2$ (see [48, p. 124]). The next result deals with the case when $s = 2$.

Lemma 5.1. *Let $F/\mathbb{F}_q$ be a function field and let E_1/F and E_2/F be distinct Galois extensions of F such that $[E_1 : F] = [E_2 : F] = p = \mathrm{char}(\mathbb{F}_q)$. Denote by $E = E_1 \cdot E_2$ the composite field of E_1 and E_2. Let Q be a place of E and denote by Q_1, Q_2 and P its restrictions to the subfields E_1, E_2 and F. Suppose that we have $d(Q_i|P) = 2 \cdot (e(Q_i|P) - 1)$ for $i = 1, 2$. Then we also have*

$$d(Q|Q_i) = 2 \cdot (e(Q|Q_i) - 1) \quad \text{for } i = 1, 2.$$

Proof. Denote by v_P the discrete valuation of F corresponding to the place P (and by v_Q, v_{Q_i} the valuations of E, E_i accordingly). The only non-trivial case of Lemma 5.1 is when $d(Q_1|P) = d(Q_2|P) = 2 \cdot (p - 1)$. By the theory of Artin-Schreier extensions (see [48, p. 115]) we can find elements $x_1 \in E_1$ and $x_2 \in E_2$ such that $E_1 = F(x_1), E_2 = F(x_2)$ and moreover

$$x_1^p - x_1 = z_1 \quad \text{and} \quad x_2^p - x_2 = z_2,$$

where z_1 and z_2 are functions in F with $v_P(z_1) = v_P(z_2) = -1$. It follows from the equation $x_1^p - x_1 = z_1$ that $v_{Q_1}(x_1) = -1$. Since the residue class field of F at the place P is perfect, there are elements u and w in F satisfying

$$\frac{z_2}{z_1} = u^p + w, \quad v_P(u) = 0 \quad \text{and} \quad v_P(w) \geq 1.$$

It follows that

$$x_2^p - x_2 = z_1 u^p + z_1 w = (x_1^p - x_1) u^p + z_1 w = ((x_1 u)^p - x_1 u) + x_1 (u - u^p) + z_1 w.$$

Setting $x_3 := x_2 - x_1 u$, we obtain that $E = E_1(x_3)$ and

$$x_3^p - x_3 = \tilde{u} x_1 + \tilde{w} =: z_3,$$

with $v_{Q_1}(\tilde{u}) \geq 0, v_{Q_1}(\tilde{w}) \geq 0$ and $v_{Q_1}(x_1) = -1$. Hence we have $v_{Q_1}(z_3) = -1$ or $v_{Q_1}(z_3) \geq 0$. If $v_{Q_1}(z_3) = -1$, then $e(Q|Q_1) = p$ and $d(Q|Q_1) = 2(p - 1)$; if $v_{Q_1}(z_3) \geq 0$, then $e(Q|Q_1) = 1$ and $d(Q|Q_1) = 0$. This shows that we have $d(Q|Q_1) = 2(e(Q|Q_1) - 1)$. $\qquad\square$

In the subsequent subsections 5.1 - 5.5 we shall investigate wild towers that were introduced in the papers [8, 9, 20, 21, 30]. Using Theorem 3.8 and Lemma 5.1, we will give simpler and less computational proofs of the asymptotic behaviour of the genus of the towers in [20, 21, 30].

5.1 The Tower $\mathcal{W}_1$

This is an asymptotically optimal wild tower over a finite field with square cardinality (see [21]). Let $q = \ell^2$ where ℓ is a power of a prime number p. We consider the tower $\mathcal{W}_1$ over $\mathbb{F}_q$ which is recursively given by the equation

$$Y^\ell + Y = \frac{X^\ell}{X^{\ell-1} + 1}. \tag{5.1}$$

First we have to show that Equation (5.1) actually defines a tower.

Lemma 5.2. *Let* $\mathcal{W}_1 = (F_0, F_1, F_2, \ldots)$ *where* $F_0 = \mathbb{F}_q(x_0)$ *is the rational function field and* $F_{n+1} = F_n(x_{n+1})$ *with the relation* $x_{n+1}^\ell + x_{n+1} = x_n^\ell/(x_n^{\ell-1} + 1)$, *for each* $n \geq 0$. *Then the following holds:*

i) *For each* $n \geq 0$, *the extension* F_{n+1}/F_n *has degree* $[F_{n+1} : F_n] = \ell$, *and the field* $\mathbb{F}_q$ *is algebraically closed in* F_{n+1}.

ii) *The pole* P_∞ *of* x_0 *in* F_0 *is totally ramified in all extensions* F_{n+1}/F_0. *If* Q_{n+1} *denotes the unique place of* F_{n+1} *above* P_∞, *then* Q_{n+1} *is a simple pole of the function* x_{n+1}.

Proof. We proceed by induction. Let Q_n be the unique place of F_n above P_∞, and let v_{Q_n} denote the corresponding valuation of F_n. By induction hypothesis, we have $v_{Q_n}(x_n) = -1$. Let Q_{n+1} be a place of F_{n+1} above Q_n. It follows from Equation (5.1) that

$$v_{Q_{n+1}}(x_{n+1}^\ell + x_{n+1}) = e(Q_{n+1}|Q_n) \cdot v_{Q_n}(x_n^\ell/(x_n^{\ell-1} + 1)) = -e(Q_{n+1}|Q_n).$$

Therefore the place Q_{n+1} is a pole of the function x_{n+1}, and

$$-e(Q_{n+1}|Q_n) = v_{Q_{n+1}}(x_{n+1}^\ell + x_{n+1}) = \ell \cdot v_{Q_{n+1}}(x_{n+1}).$$

We then conclude that $e(Q_{n+1}|Q_n) = \ell$ and that $v_{Q_{n+1}}(x_{n+1}) = -1$. All assertions of Lemma 5.2 follow now immediately. $\square$

Next we determine the ramification locus $V(\mathcal{W}_1)$ of the tower $\mathcal{W}_1$. Recall that $(x_0 = \alpha)$ denotes the place of the rational function field F_0 which is the zero of the function $x_0 - \alpha$.

Lemma 5.3. *The ramification locus* $V(\mathcal{W}_1)$ *of the tower* $\mathcal{W}_1$ *is given by*

$$V(\mathcal{W}_1) = \{P_\infty\} \cup \{(x_0 = \alpha) \mid \alpha^\ell + \alpha = 0\}.$$

Proof. Let $F = \mathbb{F}_q(x, y)$ with $y^\ell + y = x^\ell/(x^{\ell-1} + 1)$ be the basic function field of the tower $\mathcal{W}_1$. It is clear from the theory of Artin-Schreier extensions (see [48, p.115]) that exactly the places $(x = \alpha)$, with $\alpha^{\ell-1} + 1 = 0$, and the pole of x are ramified in the extension $F/\mathbb{F}_q(x)$. With notations as in Proposition 3.10, we thus have

$$R_0 = \{\infty\} \cup \{\alpha \mid \alpha^{\ell-1} + 1 = 0\}.$$

Consider now the set

$$R := \{\infty\} \cup \{\alpha \mid \alpha^{\ell} + \alpha = 0\}.$$

We want to apply Proposition 3.10 and therefore we have to show: if $\beta \in R$ and $\alpha \in \overline{\mathbb{F}}_q \cup \{\infty\}$ satisfy the equation

$$\beta^{\ell} + \beta = \alpha^{\ell}/(\alpha^{\ell-1} + 1), \tag{5.2}$$

then $\alpha \in R$. First consider the case $\beta = \infty$. It follows from Equation (5.2) that $\alpha = \infty$ or $\alpha^{\ell-1} + 1 = 0$, hence $\alpha \in R$. In case $\beta \in R \setminus \{\infty\}$, Equation (5.2) gives

$$\alpha^{\ell}/(\alpha^{\ell-1} + 1) = \beta^{\ell} + \beta = 0,$$

and it follows that $\alpha = 0 \in R$. So the hypothesis of Proposition 3.10 is satisfied, and we conclude that

$$\begin{aligned} V(\mathcal{W}_1) &\subseteq \{P \mid P \text{ is a place of } F_0 \text{ with } x_0(P) \in R\} \\ &= \{P_{\infty}\} \cup \{(x_0 = \alpha) \mid \alpha^{\ell} + \alpha = 0\}. \end{aligned}$$

The places P_{∞} and $(x_0 = \alpha)$ with $\alpha^{\ell-1} + 1 = 0$ are ramified in the extension F_1/F_0, and it is easily verified that the place $(x_0 = 0)$ ramifies in the extension F_2/F_0. This finishes the proof of Lemma 5.3. $\qquad\square$

The splitting locus $Z(\mathcal{W}_1)$ of the tower $\mathcal{W}_1$ can be easily determined by using Proposition 3.13.

Lemma 5.4. *The splitting locus $Z(\mathcal{W}_1)$ of the tower $\mathcal{W}_1$ is given by*

$$Z(\mathcal{W}_1) = \{(x_0 = \alpha) \mid \alpha \in \mathbb{F}_q \text{ and } \alpha^{\ell} + \alpha \neq 0\}.$$

In particular, the splitting rate $\nu(\mathcal{W}_1)$ satisfies the inequality $\nu(\mathcal{W}_1) \geq \ell^2 - \ell$.

Proof. Observe that the map $\gamma \mapsto \gamma^{\ell} + \gamma$ (resp. the map $\gamma \mapsto \gamma^{\ell+1}$) is the trace (resp. the norm) map from $\mathbb{F}_q = \mathbb{F}_{\ell^2}$ to the subfield $\mathbb{F}_{\ell}$. Using notations as in Proposition 3.13 we consider the set

$$S := \{\alpha \in \mathbb{F}_q \mid \alpha^{\ell} + \alpha \neq 0\}.$$

For $\alpha \in S$, the equation

$$\beta^{\ell} + \beta = \alpha^{\ell}/(\alpha^{\ell-1} + 1) = \alpha^{\ell+1}/(\alpha^{\ell} + \alpha) \in \mathbb{F}_{\ell} \setminus \{0\}$$

has ℓ distinct roots β in S. Thus Condition c) (just after the proof of Proposition 3.13) is satisfied, and it follows from Proposition 3.13 that

$$Z(\mathcal{W}_1) \supseteq \{(x_0 = \alpha) \mid \alpha \in \mathbb{F}_q \text{ and } \alpha^\ell + \alpha \neq 0\}. \tag{5.3}$$

As $V(\mathcal{W}_1) = \{P_\infty\} \cup \{(x_0 = \alpha) \mid \alpha^\ell + \alpha = 0\}$ (by Lemma 5.3) and since we have $V(\mathcal{W}_1) \cap Z(\mathcal{W}_1) = \emptyset$, we conclude that equality holds in (5.3). $\qquad\square$

The previous lemmas show that the tower $\mathcal{W}_1$ has a finite ramification locus and a positive splitting rate, so it is a promising candidate for being asymptotically good. In the next lemmas we determine the different exponents of the ramified places in the tower.

Lemma 5.5. *Let $F = \mathbb{F}_q(x, y)$ with $y^\ell + y = x^\ell/(x^{\ell-1} + 1)$ be the basic function field of the tower $\mathcal{W}_1$. Then the following holds:*

i) *Both extensions $F/\mathbb{F}_q(x)$ and $F/\mathbb{F}_q(y)$ are abelian extensions with degrees $[F : \mathbb{F}_q(x)] = [F : \mathbb{F}_q(y)] = \ell$.*

ii) *Let P be a place of $\mathbb{F}_q(x)$ (or of $\mathbb{F}_q(y)$) which is ramified in F. Then P is totally ramified in F. If Q is the place of F above P, then the different exponent of $Q|P$ is $d(Q|P) = 2(\ell - 1) = 2(e(Q|P) - 1)$.*

iii) *Let $\ell = p^a$ where $p = \operatorname{char}(\mathbb{F}_q)$ and $a \geq 1$. Then there exist intermediate fields in the extensions $F/\mathbb{F}_q(x)$ and $F/\mathbb{F}_q(y)$*

$$\mathbb{F}_q(x) = M_0 \subseteq \ldots \subseteq M_a = F \quad \text{and} \quad \mathbb{F}_q(y) = L_0 \subseteq \ldots \subseteq L_a = F$$

with the following properties:

a) *All extensions M_{i+1}/M_i and L_{i+1}/L_i are Galois of degree p.*

b) *If P_i is any place of the field M_i (resp. of L_i) and P_{i+1} is a place of M_{i+1} (resp. of L_{i+1}) above P_i, then the different exponent of $P_{i+1}|P_i$ is given by $d(P_{i+1}|P_i) = 2 \cdot (e(P_{i+1}|P_i) - 1)$.*

Proof. i) All solutions of the equation $\gamma^\ell + \gamma = 0$ are in $\mathbb{F}_q$, and the maps $y \mapsto y + \gamma$ yield ℓ distinct automorphisms of F over $\mathbb{F}_q(x)$. Hence the extension $F/\mathbb{F}_q(x)$ is Galois of degree ℓ, and the Galois group is abelian. The irreducible equation between x and y can be rewritten as

$$\left(\frac{1}{x}\right)^\ell + \frac{1}{x} = \frac{1}{y^\ell + y} \tag{5.4}$$

and the same argument as above shows that the extension $F/\mathbb{F}_q(y)$ is also abelian of degree ℓ.

ii) For the extension $F/\mathbb{F}_q(x)$, the assertion follows from the theory of Artin-Schreier extensions [48] and from the defining equation $y^\ell + y = x^\ell/(x^{\ell-1} + 1)$;

for the extension $F/\mathbb{F}_q(y)$ one considers Equation (5.4).

iii) The existence of intermediate fields M_i (resp. L_i) with Property a) is clear, from Galois theory. Property b) follows by induction from item ii) and from the following claim:

Claim: Let N/L be an abelian extension of function fields over $\mathbb{F}_q$ with degree satisfying $[N : L] = p^c$, where $p = \mathrm{char}(\mathbb{F}_q)$ and $c \geq 1$, and let Q be a place of N. Let M be an intermediate field and set $P := Q \cap L$ and $R := Q \cap M$. Suppose that $d(Q|P) = 2 \cdot (e(Q|P) - 1)$. Then it follows that

$$d(Q|R) = 2 \cdot (e(Q|R) - 1) \quad \text{and} \quad d(R|P) = 2 \cdot (e(R|P) - 1).$$

Proof of the Claim: Hilbert's different formula (see [48, p. 124]) implies that

$$d(Q|R) \geq 2 \cdot (e(Q|R) - 1) \quad \text{and} \quad d(R|P) \geq 2 \cdot (e(R|P) - 1).$$

The transitivity of different exponents then yields

$$\begin{aligned} d(Q|P) &= e(Q|R) \cdot d(R|P) + d(Q|R) \\ &\geq e(Q|R) \cdot 2 \cdot (e(R|P) - 1) + 2 \cdot (e(Q|R) - 1) \qquad (5.5) \\ &= 2 \cdot (e(Q|P) - 1) = d(Q|P). \end{aligned}$$

Hence the inequality in (5.5) must be an equality, and the claim is proved. This finishes also the proof of Lemma 5.5. □

Lemma 5.6. *Let $\mathcal{W}_1 = (F_0, F_1, F_2, \ldots)$ be the tower over $\mathbb{F}_q$ which is recursively defined by Equation (5.1). For $n \geq 1$, let Q be a place of F_n and set $P = Q \cap F_0$. Then*

$$d(Q|P) = 2 \cdot (e(Q|P) - 1).$$

Proof. We refine the pyramid associated to the tower $\mathcal{W}_1$ with intermediate fields of degree p, according to Lemma 5.5 iii). Figure 5.1 below illustrates this refinement process:

For each extension of degree p on the bottom edges $\mathbb{F}_q(x_n, x_{n+1})/\mathbb{F}_q(x_n)$ or $\mathbb{F}_q(x_n, x_{n+1})/\mathbb{F}_q(x_{n+1})$ of this pyramid, the different exponents are either $2p - 2$ or 0, as follows from Lemma 5.5. Climbing up the refined pyramid (using Lemma 5.1 repeatedly) it follows that for any subextension of degree p in the refined pyramid, the different exponents are either $2p - 2$ or 0. This holds in particular along the left edge of the pyramid which represents the tower $\mathcal{W}_1$. By the transitivity of different exponents, Lemma 5.6 now follows.

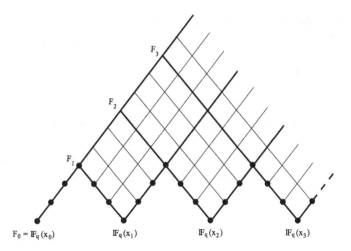

Figure 5.1

□

Theorem 5.7. *Let* $q = \ell^2$ *be a square, and let* $\mathcal{W}_1 = (F_0, F_1, F_2, \ldots)$ *be the tower over* $\mathbb{F}_q$ *which is recursively defined by the equation* $Y^\ell + Y = X^\ell/(X^{\ell-1} + 1)$. *Then we have:*

$$\gamma(\mathcal{W}_1) = \ell, \ \ \nu(\mathcal{W}_1) = \ell^2 - \ell \ and \ \lambda(\mathcal{W}_1) = \ell - 1.$$

In particular, the tower $\mathcal{W}_1$ *attains the Drinfeld-Vladut bound and it is therefore asymptotically optimal over* $\mathbb{F}_q$.

Proof. By Lemma 5.3, the ramification locus of the tower $\mathcal{W}_1$ has cardinality $\#V(\mathcal{W}_1) = \ell + 1$. From Lemma 5.6 we have that for all places $P \in V(\mathcal{W}_1)$, all $n \geq 1$ and all places Q of F_n above P, the different exponent of $Q|P$ satisfies $d(Q|P) \leq 2e(Q|P)$. Now Theorem 3.8 gives the following upper bound for the genus $\gamma(\mathcal{W}_1)$ of the tower $\mathcal{W}_1$:

$$\gamma(\mathcal{W}_1) \leq -1 + \frac{1}{2} \cdot 2 \cdot (\ell + 1) = \ell.$$

The splitting rate $\nu(\mathcal{W}_1)$ satisfies $\nu(\mathcal{W}_1) \geq \ell^2 - \ell$, by Lemma 4.4. Therefore

$$\lambda(\mathcal{W}_1) = \nu(\mathcal{W}_1)/\gamma(\mathcal{W}_1) \geq (\ell^2 - \ell)/\ell = \ell - 1.$$

On the other hand, we have $\lambda(\mathcal{W}_1) \leq \ell - 1$ by the Drinfeld-Vladut bound (see Theorem 2.5). Hence $\lambda(\mathcal{W}_1) = \ell - 1$, $\gamma(\mathcal{W}_1) = \ell$ and $\nu(\mathcal{W}_1) = \ell^2 - \ell$. □

Remark 5.8. i) It follows from the Drinfeld-Vladut bound and from Theorem 5.7 that $A(q) = \sqrt{q} - 1$ if q is a square.

ii) A detailed analysis of the ramification behaviour of the places $P \in V(\mathcal{W}_1)$ in the tower $\mathcal{W}_1$ yields a precise formula for the genus $g(F_n)$ as follows:

$$g(F_n) = \begin{cases} (q^{(n+1)/2} - 1)^2 & \text{for } n \equiv 1 \bmod 2; \\ (q^{n/2} - 1)(q^{(n+2)/2} - 1) & \text{for } n \equiv 0 \bmod 2. \end{cases} \quad (5.6)$$

The proof of Equation (5.6) is very technical and it requires subtle "pole order reductions", see [21]. If one just settles for the asymptotic result (i.e., for the genus $\gamma(\mathcal{W}_1) = \ell$ of the tower $\mathcal{W}_1$), the proof of Theorem 5.7 given here (following the paper [26]) is much easier and shorter than the original one given in [21].

5.2   The Tower $\mathcal{W}_2$

As in Section 5.1 let $q = \ell^2$ be a square. For each divisor $m \geq 1$ of $(\ell + 1)$, we will construct here an asymptotically optimal tower $\mathcal{W}_2$ over $\mathbb{F}_q$. In the special case $m = \ell + 1$, the tower $\mathcal{W}_2$ coincides with the first explicit example in the literature attaining the Drinfeld-Vladut bound, which is the content of the paper [20].

We start with the tower $\mathcal{E} = (E_0, E_1, E_2, \ldots)$ over $\mathbb{F}_q$ which is defined as follows:

$E_0 = \mathbb{F}_q(z_0)$ is the rational function field,

$E_1 = E_0(z_1) = \mathbb{F}_q(z_1)$ with $z_1^\ell + z_1 = z_0$, and $\quad (5.7)$

$E_{n+1} = E_n(z_{n+1})$ with $z_{n+1}^\ell + z_{n+1} = z_n^\ell/(z_n^{\ell-1} + 1)$, for all $n \geq 1$.

It is clear that the tower $(E_1, E_2, E_3, \ldots)$ is just the tower $\mathcal{W}_1$ that was discussed in Section 5.1, and that $E_0 \subseteq E_1$ is the subfield $E_0 = \mathbb{F}_q(z_1^\ell + z_1) \subseteq \mathbb{F}_q(z_1) = E_1$. The place $(z_0 = 0)$ of E_0 splits completely in E_1/E_0 into the places $(z_1 = \alpha)$ of E_1 with $\alpha^\ell + \alpha = 0$. The place $(z_0 = \infty)$ is the only ramified place in the extension E_1/E_0; its extension to E_1 is the pole $(z_1 = \infty)$ of z_1, and the different exponent d of $(z_1 = \infty)$ over $(z_0 = \infty)$ is given by $d = 2(\ell - 1)$. Therefore the ramification locus of the tower $\mathcal{E}$ is (see Lemma 5.3)

$$V(\mathcal{E}) = \{(z_0 = 0), (z_0 = \infty)\}.$$

For a place $P_0 \in V(\mathcal{E})$, an integer $n \geq 1$ and a place Q_0 of the field E_n lying above P_0, we have

$$d(Q_0|P_0) = 2(e(Q_0|P_0) - 1). \quad (5.8)$$

Equation (5.8) follows from Lemma 5.6 and the transitivity of different exponents applied to the extensions $E_0 \subseteq E_1 \subseteq E_n$. Now we fix an integer $m \geq 1$

which is a divisor of $(\ell + 1)$ and we consider the fields

$$F_0 := \mathbb{F}_q(x_0) \text{ with } x_0^m = z_0, \text{ and}$$

(5.9)

$$F_n := F_0 \cdot E_n \text{ for all } n \geq 1.$$

It is obvious that we thus obtain a tower

$$\mathcal{W}_2 := (F_0, F_1, F_2, \ldots) \text{ over } \mathbb{F}_q.$$

Figure 5.2 illustrates the towers $\mathcal{E}$ and $\mathcal{W}_2$:

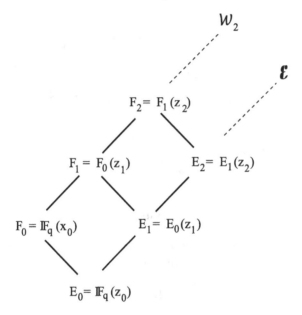

Figure 5.2

For all $i \geq 0$, the degrees of the field extensions in Figure 5.2 are

$$[F_{i+1} : F_i] = [E_{i+1} : E_i] = \ell \text{ and } [F_i : E_i] = m.$$

Theorem 5.9. *Let $q = \ell^2$ be a square and let $m \geq 1$ be a divisor of $(\ell + 1)$. Then the tower $\mathcal{W}_2$ over $\mathbb{F}_q$ which is defined by (5.7) and (5.9) is asymptotically optimal; i.e., its limit is $\lambda(\mathcal{W}_2) = \ell - 1$, and it attains therefore the Drinfeld-Vladut bound.*

Proof. We consider the two towers $\mathcal{E}$ and $\mathcal{W}_2$ as illustrated in Figure 5.2. From the equation $x_0^m = z_0$ it follows that in the extension F_0/E_0 just the two places $(z_0 = 0)$ and $(z_0 = \infty)$ are (totally and tamely) ramified; the places of F_0

above them are the places $(x_0 = 0)$ and $(x_0 = \infty)$, respectively. All other places of E_0 are unramified in F_0/E_0. Since the tower $\mathcal{E}$ has the ramification locus $V(\mathcal{E}) = \{(z_0 = 0), (z_0 = \infty)\}$, we conclude that the tower $\mathcal{W}_2$ has ramification locus

$$V(\mathcal{W}_2) = \{(x_0 = 0), (x_0 = \infty)\}.$$

Let $P \in V(\mathcal{W}_2)$ and let Q be a place of F_n lying above P. Set $P_0 = P \cap E_0$, $Q_0 = Q \cap E_n$ and $e = e(Q_0|P_0)$. Since the ramification index $e(Q_0|P_0)$ is a power of the characteristic p and since m is a divisor of $(\ell + 1)$, it follows from Abhyankar's lemma that we have the following situation:

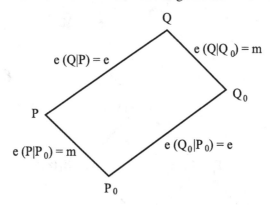

Figure 5.3

We now calculate the different exponent $d(Q|P_0)$ in two different ways, using the transitivity of different exponents:

$$d(Q|P_0) = e \cdot d(P|P_0) + d(Q|P) = e \cdot (m - 1) + d(Q|P),$$

$$d(Q|P_0) = m \cdot d(Q_0|P_0) + d(Q|Q_0) = m \cdot (2e - 2) + (m - 1).$$

In the last equality above we have used Equation (5.8). Hence

$$d(Q|P) = m \cdot (2e - 2) + (m - 1) - e(m - 1) = (m + 1)(e - 1). \quad (5.10)$$

Now using Theorem 3.8 with $c_P = m + 1$ (see Equation (5.10)) one gets an upper bound for the genus of the tower $\mathcal{W}_2$ as follows:

$$\gamma(\mathcal{W}_2) \le -1 + \frac{1}{2} \cdot 2 \cdot (m + 1) = m. \quad (5.11)$$

Next we determine the splitting locus $Z(\mathcal{W}_2)$ of the tower $\mathcal{W}_2$. From Lemma 5.4 and from the equation $z_1^\ell + z_1 = z_0$, it is obvious that the splitting locus of the tower $\mathcal{E}$ is

$$Z(\mathcal{E}) = \{(z_0 = \alpha) \mid \alpha \in \mathbb{F}_\ell \text{ and } \alpha \ne 0\} = \{(z_0 = \alpha) \mid \alpha^{\ell-1} = 1\}.$$

For $\alpha \in \mathbb{F}_\ell \setminus \{0\}$, the equation $\beta^m = \alpha$ has m distinct roots $\beta \in \mathbb{F}_q$, since m is a divisor of $(\ell + 1)$. Therefore all places in $Z(\mathcal{E})$ split completely in F_0/E_0 and hence we have

$$Z(\mathcal{W}_2) \supseteq \{(x_0 = \beta) \mid \beta^{m(\ell-1)} = 1\}.$$

This implies that the splitting rate $\nu(\mathcal{W}_2)$ satisfies the inequality

$$\nu(\mathcal{W}_2) \geq m \cdot (\ell - 1). \tag{5.12}$$

We conclude from (5.11) and (5.12) that

$$\lambda(\mathcal{W}_2) = \nu(\mathcal{W}_2)/\gamma(\mathcal{W}_2) \geq m \cdot (\ell - 1)/m = \ell - 1,$$

hence $\lambda(\mathcal{W}_2) = \ell - 1$ and the tower $\mathcal{W}_2$ is asymptotically optimal. $\square$

Remark 5.10. We consider the special case $m = \ell + 1$ and we define for $n \geq 1$ the elements $x_n \in F_n$ recursively by

$$x_n := z_n/x_{n-1}. \tag{5.13}$$

It follows from the defining equations (5.7) and (5.9) of the tower $\mathcal{W}_2$ that the following relation holds:

$$z_{n+1}^\ell + z_{n+1} = x_n^{\ell+1}, \quad \text{for all } n \geq 0. \tag{5.14}$$

Equations (5.13) and (5.14) are just the defining equations of the asymptotically optimal tower in the paper [20]. We have thus obtained a new (and much simpler) proof of the main result of [20], which states that $\gamma(\mathcal{W}_2) = \ell + 1$ in the case $m = \ell + 1$.

5.3 The Tower $\mathcal{W}_3$

We describe now another wild tower which was investigated in detail in the paper [9]. As before, let $q = \ell^2$ be a square. Our starting point is the tower $\mathcal{W}_1$ from Section 5.1 which is given recursively by $Y^\ell + Y = X^\ell/(X^{\ell-1}+1)$. So $\mathcal{W}_1 = (L_0, L_1, L_2, \dots)$ where $L_0 = \mathbb{F}_q(t_0)$ and for each $n \geq 0$, $L_{n+1} = L_n(t_{n+1})$ with the relation

$$t_{n+1}^\ell + t_{n+1} = t_n^\ell/(t_n^{\ell-1} + 1). \tag{5.15}$$

Defining the function $x_n \in L_n$ by $x_n := (t_n^{\ell-1} + 1)^{-1}$, a straightforward calculation shows that the equation below holds (for all $n \geq 0$):

$$\frac{x_{n+1} - 1}{x_{n+1}^\ell} = \frac{x_n^\ell - 1}{x_n}. \tag{5.16}$$

We set $F_n := \mathbb{F}_q(x_0, x_1, \ldots, x_n)$ and thus obtain a subtower $\mathcal{W}_3 = (F_0, F_1, \ldots)$ of the tower $\mathcal{W}_1$. As follows from Proposition 3.6, the tower $\mathcal{W}_3$ is also an asymptotically optimal tower over $\mathbb{F}_q$. We summarize these results in the next theorem:

Theorem 5.11. *Let $q = \ell^2$ be a square. Then the equation*

$$\frac{Y - 1}{Y^\ell} = \frac{X^\ell - 1}{X} \tag{5.17}$$

defines recursively a tower $\mathcal{W}_3 = (F_0, F_1, F_2, \ldots)$ over $\mathbb{F}_q$ with $\lambda(\mathcal{W}_3) = \ell - 1$; i.e., the tower $\mathcal{W}_3$ attains the Drinfeld-Vladut bound.

Remark 5.12. i) The tower $\mathcal{W}_3$ has some interesting features. While in the towers $\mathcal{W}_1$ and $\mathcal{W}_2$ all steps F_{i+1}/F_i are Galois, this is not true for the tower $\mathcal{W}_3$ (if $\ell \neq 2$). Also, if $\ell \neq 2$ there occurs both wild and tame ramification in the tower $\mathcal{W}_3$. As before (see Remark 5.8) an explicit formula for the genus $g(F_n)$ requires subtle "pole order reductions" (see [9] for details).
ii) One can show that the tower $\mathcal{W}_3$ is the same as the tower in [16, Equation (25)], where it is proved that the towers $\mathcal{W}_1, \mathcal{W}_2$ and $\mathcal{W}_3$ above are Drinfeld modular.

5.4   The Tower $\mathcal{W}_4$

It seems to be more difficult to exhibit asymptotically good wild towers over finite fields with non-square cardinalities. The first explicit example was found by van der Geer - van der Vlugt [30]; it is a tower over the field with eight elements. We present now their example, giving a simplified proof for the limit of the tower. The van der Geer - van der Vlugt tower $\mathcal{W}_4 = (F_0, F_1, F_2, \ldots)$ over the field $\mathbb{F}_8$ is recursively defined by the equation

$$Y^2 + Y = X + 1 + 1/X. \tag{5.18}$$

Lemma 5.13. *Equation (5.18) defines a tower $\mathcal{W}_4$ over the field $\mathbb{F}_8$. We have $[F_n : F_0] = 2^n$ for all $n \geq 0$, and the pole P_∞ of the function x_0 is totally ramified in all extensions F_n/F_0.*

Proof. Similar to the proof of Lemma 5.2. $\qquad\qquad\qquad\qquad\square$

We want to apply Proposition 3.13 to determine the splitting locus $Z(\mathcal{W}_4)$ of the tower $\mathcal{W}_4$.

Lemma 5.14. *We have $Z(\mathcal{W}_4) \supseteq \{(x_0 = \alpha) \mid \alpha \in \mathbb{F}_8 \setminus \mathbb{F}_2\}$. Hence the splitting rate satisfies $\nu(\mathcal{W}_4) \geq 6$.*

Proof. Consider the basic function field $F = \mathbb{F}_8(x, y)$ with $y^2 + y = x + 1 + 1/x$ and the set

$$S := \mathbb{F}_8 \setminus \mathbb{F}_2 = \{\alpha \in \bar{\mathbb{F}}_8 \mid \alpha^6 + \alpha^5 + \alpha^4 + \alpha^3 + \alpha^2 + \alpha + 1 = 0\}.$$

Let $\alpha \in S$ and consider the solutions $\beta \in \overline{\mathbb{F}}_8$ of the equation $\beta^2 + \beta = \alpha + 1 + \alpha^{-1}$. Then one has

$$\beta^2 + \beta = \alpha + 1 + \alpha^{-1}, \beta^4 + \beta^2 = \alpha^2 + 1 + \alpha^{-2} \text{ and } \beta^8 + \beta^4 = \alpha^4 + 1 + \alpha^{-4}.$$

Adding these equations we obtain

$$\begin{aligned}
\beta^8 + \beta &= \alpha^4 + \alpha^2 + \alpha + 1 + \alpha^{-1} + \alpha^{-2} + \alpha^{-4} \\
&= \alpha^{-4}(\alpha^8 + \alpha^6 + \alpha^5 + \alpha^4 + \alpha^3 + \alpha^2 + 1) \\
&= \alpha^{-4}(\alpha^6 + \alpha^5 + \alpha^4 + \alpha^3 + \alpha^2 + \alpha + 1) = 0.
\end{aligned}$$

Hence $\beta \in \mathbb{F}_8$, so Condition a) of Proposition 3.13 is satisfied. It is also clear that $\beta \neq 0, 1$, since otherwise $\alpha + 1 + \alpha^{-1} = 0$ and then $\alpha \notin \mathbb{F}_8$. So we have $\beta \in S$ and also Condition b) of Proposition 3.13 holds. Now Lemma 5.14 follows from Proposition 3.13. □

Next we determine the ramification locus $V(\mathcal{W}_4)$ of the tower $\mathcal{W}_4$.

Lemma 5.15. *We have $V(\mathcal{W}_4) \subseteq \{(x_0 = \alpha) \mid \alpha \in \mathbb{F}_4 \text{ or } \alpha = \infty\}$.*

Proof. We use here Proposition 3.10. Let $F = \mathbb{F}_8(x, y)$ be the basic function field of the tower $\mathcal{W}_4$ with $y^2 + y = x + 1 + 1/x$. Only the places $(x = 0)$ and $(x = \infty)$ are ramified in the extension $F/\mathbb{F}_8(x)$, so the set R_0 as defined in (3.2) is $R_0 = \{0, \infty\}$. Let $\beta \in R := \mathbb{F}_4 \cup \{\infty\}$, and let $\alpha \in \overline{\mathbb{F}}_8 \cup \{\infty\}$ be a solution of the equation

$$\beta^2 + \beta = \alpha + 1 + \alpha^{-1}.$$

If $\beta = \infty$, then $\alpha = 0$ or $\alpha = \infty$. If $\beta \in \mathbb{F}_4$, then $\beta^2 + \beta \in \mathbb{F}_2$ and α satisfies an equation of degree 2 over $\mathbb{F}_2$, hence $\alpha \in \mathbb{F}_4$. In all cases we have proved that $\alpha \in R$. Now the assertion of Lemma 5.15 follows from Proposition 3.10. □

In fact it is an easy exercise to prove that the ramification locus $V(\mathcal{W}_4)$ is equal to the set $\{(x_0 = \alpha) \mid \alpha \in \mathbb{F}_4 \text{ or } \alpha = \infty\}$: for example, there is a place P of F_2 such that $x_0(P) = 1$, $x_1(P) \in \mathbb{F}_4 \setminus \mathbb{F}_2$ and $x_2(P) = 0$; this place is then ramified in the extension F_3/F_2.

We also have an analogue to Lemma 5.5 for the basic function field associated to the tower $\mathcal{W}_4$:

Lemma 5.16. *Let $F = \mathbb{F}_8(x, y)$ with $y^2 + y = x + 1 + 1/x$ be the basic function field of the tower $\mathcal{W}_4$. Then both extensions $F/\mathbb{F}_8(x)$ and $F/\mathbb{F}_8(y)$ are Galois of degree 2. If P is a place of $\mathbb{F}_8(x)$ (or of $\mathbb{F}_8(y)$) which is ramified in F, and if Q is the place of F lying above P, then the different exponent of $Q|P$ is $d(Q|P) = 2$.*

Proof. This follows immediately from the two equations below (see [48, p.115]):

$$y^2 + y = x + 1 + 1/x \quad \text{and} \quad \left(\frac{x}{x+1}\right)^2 + \left(\frac{x}{x+1}\right) = \frac{1}{y^2+y+1}.$$

$\square$

Theorem 5.17. *The tower $\mathcal{W}_4$ over $\mathbb{F}_8$, which is recursively defined by the equation $Y^2 + Y = X + 1 + 1/X$, is asymptotically good. Its limit $\lambda(\mathcal{W}_4)$ satisfies*

$$\lambda(\mathcal{W}_4) \geq 3/2.$$

It follows in particular that Ihara's quantity $A(q)$ for $q = 8$ satisfies the inequality $A(8) \geq 3/2$.

Proof. Let $\mathcal{W}_4 = (F_0, F_1, F_2, \ldots)$, let Q be a place of F_n and let $P = Q \cap F_0$ be the place of F_0 below Q. As in Lemma 5.6, it follows from Lemma 5.16 that the different exponent of $Q|P$ is $d(Q|P) = 2(e(Q|P) - 1)$. Now Theorem 3.8 with $c_P = 2$ and Lemma 5.15 yield

$$\gamma(\mathcal{W}_4) \leq -1 + \frac{1}{2} \cdot 5 \cdot 2 = 4.$$

By Lemma 4.14 we have $\nu(\mathcal{W}_4) \geq 6$ and hence

$$\lambda(\mathcal{W}_4) = \nu(\mathcal{W}_4)/\gamma(\mathcal{W}_4) \geq 6/4 = 3/2.$$

$\square$

Remark 5.18. i) One can calculate precisely the genus of all fields F_n in the tower $\mathcal{W}_4$. These calculations are long and very technical, see [30]. For instance, one obtains the formula

$$g(F_n) = 2^{n+1} + 1 - (n + 2 \cdot [n/4] + 15) \cdot 2^{(n-3)/2}$$

for $n \equiv 1 \bmod 2$.

ii) For p a prime number, one has the following lower bound for the quantity $A(p^3)$ which is due to Zink (see [61]):

$$A(p^3) \geq 2(p^2 - 1)/(p + 2).$$

So the tower $\mathcal{W}_4$ gives a proof of this bound in the particular case $p = 2$.

5.5   The Tower $\mathcal{W}_5$

Our last example of an asymptotically good wild tower is a generalization of the van der Geer - van der Vlugt tower presented in Section 5.4. It was studied in detail in [8]. The constant field $\mathbb{F}_q$ has now cubic cardinality; i.e.,

$$q = \ell^3 \text{ for some prime power } \ell.$$

We define the tower $\mathcal{W}_5$ over $\mathbb{F}_q$ recursively by the equation

$$Y^\ell - Y^{\ell-1} = 1 - X + X^{-(\ell-1)}. \qquad (5.19)$$

For $\ell = 2$ this equation is the same as the defining equation (see Equation (5.18)) of the tower $\mathcal{W}_4$. The tower in [8] is given recursively by

$$\frac{1-V}{V^\ell} = \frac{U^\ell + U - 1}{U}. \qquad (5.20)$$

The change of variables $X = U^{-1}$ and $Y = V^{-1}$ shows that Equation (5.19) and Equation (5.20) define in fact the same tower.

It is easily seen that Equation (5.19) defines a tower; the proof is similar to the proof of Lemma 5.2. In order to determine the splitting locus $Z(\mathcal{W}_5)$ we rewrite Equation (5.19) as follows:

$$f(Y) = 1 - \frac{1}{f\left(\dfrac{X}{X-1}\right)} \text{ with } f(T) := T^\ell - T^{\ell-1}. \qquad (5.21)$$

One verifies easily that the polynomial $f(T)$ satisfies the equation

$$T \cdot (f(T)^{\ell+1} - f(T) + 1) = (T-1)^{\ell^2+\ell+1} + 1. \qquad (5.22)$$

We are going to use Proposition 3.13 and therefore we consider the set

$$S := \{\alpha \in \overline{\mathbb{F}}_q \mid (\alpha-1)^{\ell^2+\ell+1} = -1\} \setminus \{0\}. \qquad (5.23)$$

Since $q = \ell^3$, it is clear that $S \subseteq \mathbb{F}_q$. From Equation (5.22) it follows that

$$S = \{\alpha \in \overline{\mathbb{F}}_q \mid f(\alpha)^{\ell+1} - f(\alpha) + 1 = 0\}. \qquad (5.24)$$

Using (5.23) one checks that

$$\alpha \in S \implies \frac{\alpha}{\alpha-1} \in S. \qquad (5.25)$$

Take now an element $\alpha \in S$ and let $\beta \in \overline{\mathbb{F}}_q$ be such that

$$\beta^\ell - \beta^{\ell-1} = 1 - \alpha + \alpha^{-(\ell-1)}.$$

Setting $\delta := \alpha/(\alpha - 1)$ we then have

$$f(\beta) = 1 - \frac{1}{f(\delta)} \quad \text{with} \ \delta \in S,$$

by (5.21) and (5.25). Therefore from Equation (5.24) we have

$$f(\beta)^{\ell+1} - f(\beta) + 1 = \left(1 - \frac{1}{f(\delta)}\right)^{\ell+1} + \frac{1}{f(\delta)}$$

$$= \frac{1}{f(\delta)^{\ell+1}}(f(\delta)^{\ell+1} - f(\delta) + 1) = 0.$$

It follows that $\beta \in S$, and we conclude from Proposition 3.13:

Lemma 5.19. *All places* $(x_0 = \alpha)$ *with* $(\alpha - 1)^{\ell^2+\ell+1} = -1$ *and* $\alpha \neq 0$
split completely in the tower $\mathcal{W}_5$. *In particular we have that the splitting rate
satisfies*

$$\nu(\mathcal{W}_5) \geq \ell^2 + \ell.$$

It is much more difficult to determine the genus $\gamma(\mathcal{W}_5)$ of the tower $\mathcal{W}_5$. For
the proof of the following result we refer to [8].

Theorem 5.20. *Let* $q = \ell^3$ *where* ℓ *is any prime power, and let* $\mathcal{W}_5$ *be the tower
over* $\mathbb{F}_q$ *which is recursively defined by the equation*

$$Y^\ell - Y^{\ell-1} = 1 - X + X^{-(\ell-1)}.$$

Then we have

$$\gamma(\mathcal{W}_5) = \frac{\ell^2 + 2\ell}{2(\ell - 1)}, \ \nu(\mathcal{W}_5) \geq \ell^2 + \ell \ \text{and} \ \lambda(\mathcal{W}_5) \geq \frac{2(\ell^2 - 1)}{\ell + 2}.$$

Corollary 5.21. *For any cubic prime power* $q = \ell^3$ *one has*

$$A(\ell^3) \geq \frac{2(\ell^2 - 1)}{\ell + 2}.$$

Remark 5.22. i) If $\ell = p$ is a prime number, Corollary 5.21 is Zink's lower
bound for $A(p^3)$ which was first obtained by considering degenerations of certain modular surfaces, see [61].

ii) The ramification behaviour of the tower $\mathcal{W}_5$ is rather complicated, see [8] and
[30]. For the case $\ell \neq 2$, all steps F_{n+1}/F_n in the tower $\mathcal{W}_5 = (F_0, F_1, F_2, \ldots)$
are non-Galois. In this case, some places are wildly ramified and some others
are tamely ramified. Again here the detailed analysis of the different exponents
of the ramified places gives an exact formula for the genus $g(F_n)$ of all function
fields F_n in the tower $\mathcal{W}_5$.

6. Miscellaneous Results

In this section we discuss some specific aspects of towers, and in particular the asymptotic behaviour of the genus of a tower. The results of Sections 6.1-6.3 often lead to a quick decision about asymptotical badness of a tower. In Section 6.4 we present some classification results for Artin-Schreier towers.

6.1 Genus and Different Degree

Let $\mathcal{F} = (F_0, F_1, F_2, \ldots)$ be a tower of function fields over $\mathbb{F}_q$. Then the field extensions F_{n+1}/F_n are separable of degree $[F_{n+1} : F_n] > 1$, for all $n \geq 0$. The genus $g(F_n)$ is related to the degree of the different $\mathrm{Diff}(F_n/F_0)$ by the Hurwitz genus formula. We want to explore this relation in detail. For all $n \geq 1$, we denote by D_n the degree of the different of the extension F_n/F_{n-1}:

$$D_n := \deg \mathrm{Diff}(F_n/F_{n-1}).$$

Theorem 6.1. *Let* $\mathcal{F} = (F_0, F_1, F_2, \ldots)$ *be a tower of function fields over* $\mathbb{F}_q$. *Suppose that there exists a constant* $\epsilon \in \mathbb{R}$ *with* $0 \leq \epsilon < 1$ *such that the following inequality holds:*

$$D_n \leq \epsilon \cdot [F_n : F_{n-1}] \cdot D_{n-1}, \quad \text{for all } n \geq 2. \tag{6.1}$$

Then the genus $\gamma(\mathcal{F})$ *of the tower* $\mathcal{F}$ *is finite and the estimate below holds:*

$$\gamma(\mathcal{F}) \leq g(F_0) - 1 + \frac{D_1}{2(1 - \epsilon) \cdot [F_1 : F_0]}.$$

Proof. It follows immediately from (6.1) and from the transitivity of the different that the following inequality

$$\deg \mathrm{Diff}(F_n/F_0) \leq \left(\sum_{j=0}^{n-1} \epsilon^j \right) \cdot D_1 \cdot [F_n : F_1]$$

holds, for all $n \geq 1$. Using Hurwitz genus formula, we then have:

$$
\begin{aligned}
2g(F_n) - 2 &= [F_n : F_0](2g(F_0) - 2) + \deg \mathrm{Diff}(F_n/F_0) \\
&\leq [F_n : F_0] \cdot \left(2g(F_0) - 2 + \frac{D_1}{[F_1 : F_0]} \cdot \sum_{j=0}^{n-1} \epsilon^j \right) \\
&\leq [F_n : F_0] \cdot \left(2g(F_0) - 2 + \frac{D_1}{(1 - \epsilon)[F_1 : F_0]} \right).
\end{aligned}
$$

Hence we obtain the desired estimate:

$$\gamma(\mathcal{F}) = \lim_{n\to\infty} \frac{g(F_n)}{[F_n : F_0]} \leq g(F_0) - 1 + \frac{D_1}{2(1-\epsilon) \cdot [F_1 : F_0]}.$$

$\square$

A counterpart to Theorem 6.1 is the following result.

Theorem 6.2. *Let* $\mathcal{F} = (F_0, F_1, F_2, \ldots)$ *be a tower over* $\mathbb{F}_q$. *Suppose that there exist positive real numbers* $\rho_1, \rho_2, \ldots$ *with the following properties:*

a) $\rho_n \leq D_n = \deg \mathrm{Diff}(F_n/F_{n-1})$, *for all* $n \geq 1$.

b) $\rho_n \geq [F_n : F_{n-1}] \cdot \rho_{n-1}$, *for all* $n \geq 2$.

Then the genus of the tower is infinite. In particular the tower is asymptotically bad; i.e., its limit satisfies $\lambda(\mathcal{F}) = 0$.

Proof. Using again the transitivity of the different, one shows by induction that

$$\deg \mathrm{Diff}(F_n/F_0) \geq [F_n : F_1] \cdot n \cdot \rho_1, \quad \text{for all } n \geq 1.$$

Therefore from Hurwitz genus formula, we have that

$$2g(F_n) - 2 \geq [F_n : F_0](2g(F_0) - 2) + [F_n : F_1] \cdot n \cdot \rho_1.$$

Dividing this inequality by $[F_n : F_0]$ and letting $n \to \infty$, we see that the genus of the tower satisfies $\gamma(\mathcal{F}) = \lim_{n\to\infty} g(F_n)/[F_n : F_0] = \infty$. $\square$

6.2   Skew Towers are Bad

Let $\mathcal{F}$ be a recursive tower with defining equation $f(X, Y) = 0$. Considering the examples of asymptotically good towers in Section 4 and Section 5 one notes in all cases that $\deg_X f(X, Y) = \deg_Y f(X, Y)$. This is in fact a necessary condition for the tower to be asymptotically good:

Theorem 6.3. (see [23]). *Let* $\mathcal{F} = (F_0, F_1, F_2, \ldots)$ *be a recursive tower over* $\mathbb{F}_q$ *defined by the equation* $f(X, Y) = 0$. *If* $\deg_X f(X, Y) \neq \deg_Y f(X, Y)$, *then* $\lambda(\mathcal{F}) = 0$; *i.e., the tower* $\mathcal{F}$ *is asymptotically bad.*

Proof. We set $a := \deg_Y f$ and $b := \deg_X f$. It follows from the definition of a recursive tower (see Definition 2.8) that $[F_{n+1} : F_n] = a$ and hence we have (for all $n \geq 1$):

$$[F_n : F_0] = a^n \quad \text{and} \quad [F_n : \mathbb{F}_q(x_n)] = b^n. \tag{6.2}$$

This means that the pyramid (see Figure 2.1) attached to the tower $\mathcal{F}$ is "skew". Now we distinguish the cases $a > b$ and $a < b$.

Case 1: $a > b$. Considering the extension $F_n/\mathbb{F}_q(x_n)$, we see from (6.2) that $N(F_n) \leq b^n(q+1)$. Therefore the splitting rate $\nu(\mathcal{F})$ satisfies

$$\nu(\mathcal{F}) = \lim_{n \to \infty} N(F_n)/[F_n : F_0] \leq \lim_{n \to \infty} (q+1) \cdot (b/a)^n = 0.$$

By Corollary 3.5, the tower is then asymptotically bad.

Case 2: $a < b$. Fix an index $r \geq 0$ such that $g(F_r) \geq 2$. For $n \geq r$ we consider the field $E_r := \mathbb{F}_q(x_n, x_{n-1}, \ldots, x_{n-r}) \subseteq F_n$ which is isomorphic to F_r. The Hurwitz genus formula for the extension F_n/E_r gives

$$2g(F_n) - 2 \geq (2g(E_r) - 2) \cdot [F_n : E_r] \geq 2b^{n-r},$$

hence $g(F_n) \geq b^{n-r}$. It follows that the genus $\gamma(\mathcal{F})$ satisfies

$$\gamma(\mathcal{F}) = \lim_{n \to \infty} g(F_n)/[F_n : F_0] \geq \lim_{n \to \infty} b^{n-r}/a^n = \infty.$$

By Corollary 3.5, the tower $\mathcal{F}$ is asymptotically bad. $\qquad\qquad\square$

Example 6.4. As an application of Theorem 6.3 we investigate a generalization of the tower $\mathcal{W}_1$ which was discussed in Section 5.1. Let $q = \ell^s$ where ℓ is a prime power and $s \geq 2$. Let

$$\tau(T) := T + T^\ell + \ldots + T^{\ell^{s-1}} \quad \text{and} \quad \mu(T) := T^{1+\ell+\ldots+\ell^{s-1}}.$$

The polynomial $\tau(T)$ (resp. $\mu(T)$) represents the trace map (resp. the norm map) from $\mathbb{F}_q$ to its subfield $\mathbb{F}_\ell$. We then consider the tower $\mathcal{F}$ over $\mathbb{F}_q$ which is defined recursively by the equation

$$\tau(Y) = \frac{\mu(X)}{\tau(X)}. \tag{6.3}$$

Observe that Equation (6.3) coincides with Equation (5.1) in the particular case $q = \ell^2$. As in Section 5.1 one can easily show that Equation (6.3) indeed defines a recursive tower $\mathcal{F} = (F_0, F_1, F_2, \ldots)$ over the field $\mathbb{F}_q$ with the following properties:

i) For all $n \geq 0$, the extension F_{n+1}/F_n is Galois of degree $[F_{n+1} : F_n] = \ell^{s-1}$ (cf. Lemma 5.2).

ii) All places $(x_0 = \alpha)$ with $\alpha \in \mathbb{F}_q$ and $\tau(\alpha) \neq 0$ are completely splitting in the tower $\mathcal{F}$ (cf. Lemma 5.4).

For $s = 2$, the tower is asymptotically optimal over $\mathbb{F}_q$ (see Theorem 5.7). For $s \geq 3$ however, the tower is asymptotically bad. This follows from Theorem 6.3, since the degree of the left hand side of Equation (6.3) is $\deg \tau(Y) = \ell^{s-1}$, and the right hand side $\psi(X) = \mu(X)/\tau(X)$ of Equation (6.3) has degree

$$\deg \psi(X) = \ell^{s-1} + \ell^{s-2} + \ldots + \ell \neq \deg \tau(Y).$$

The last inequality above follows since $s \geq 3$.

6.3   The Dual Tower

In this section we consider recursive towers. To such a tower $\mathcal{F}$ we shall associate another tower $\mathcal{G}$ (called the dual tower of $\mathcal{F}$), and we shall study relationships between $\mathcal{F}$ and $\mathcal{G}$. The results of this section are from [6].

Definition 6.5. Let $\mathcal{F}$ be a recursive tower over $\mathbb{F}_q$ which is defined by the polynomial $f(X, Y) \in \mathbb{F}_q[X, Y]$. Then its *dual tower* $\mathcal{G}$ is the tower which is defined recursively by the polynomial $f(Y, X)$.

Definition 6.5 means that $\mathcal{F} = (F_0, F_1, F_2, \dots)$ and $\mathcal{G} = (G_0, G_1, G_2, \dots)$ where

$$F_n = \mathbb{F}_q(x_0, x_1, \dots, x_n) \text{ with } f(x_i, x_{i+1}) = 0,$$
$$G_n = \mathbb{F}_q(y_0, y_1, \dots, y_n) \text{ with } f(y_{i+1}, y_i) = 0,$$

for all $n \geq 1$ and all indices i with $0 \leq i \leq n - 1$. We identify the rational function fields $F_0 = \mathbb{F}_q(x_0)$ and $G_0 = \mathbb{F}_q(y_0)$ by setting $x_0 = y_0$, and then we have the following picture:

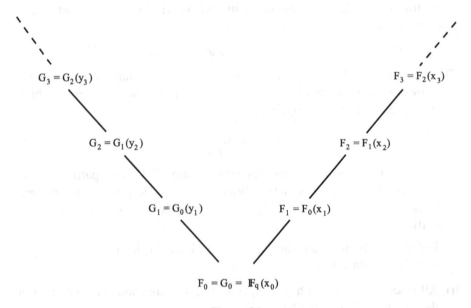

Figure 6.1

It is clear that the function fields F_n and G_n are isomorphic, for all $n \geq 0$; hence $N(F_n) = N(G_n)$ and $g(F_n) = g(G_n)$. It follows that the towers $\mathcal{F}$ and $\mathcal{G}$ have the same limit $\lambda(\mathcal{F}) = \lambda(\mathcal{G})$. Moreover, assuming that $\deg_X f(X, Y) = \deg_Y f(X, Y)$, which is a natural assumption (see Theorem 6.3), then one also has

$$\gamma(\mathcal{F}) = \gamma(\mathcal{G}) \text{ and } \nu(\mathcal{F}) = \nu(\mathcal{G}).$$

Recall that the ramification locus of the tower $\mathcal{F}$ is defined as (see Definition 3.7)

$$V(\mathcal{F}) = \{P \mid P \text{ is a place of } F_0 \text{ which ramifies in } \mathcal{F}\}.$$

We also define the *wild ramification locus* $V_w(\mathcal{F})$ as follows:

$$V_w(\mathcal{F}) := \{P \mid P \text{ is a place of } F_0 \text{ which is wildly ramified in } \mathcal{F}\}.$$

So a place P of F_0 belongs to $V_w(\mathcal{F})$ if and only if there is some $n \geq 1$ and some place Q of F_n lying above P such that the ramification index $e(Q|P)$ is divisible by the characteristic of $\mathbb{F}_q$.

Theorem 6.6. (see [6, 57]). *Suppose that $\mathcal{F}$ is a recursive tower, given recursively by the polynomial $f(X,Y)$. Let $\mathcal{G}$ be the dual tower of $\mathcal{F}$ and assume that $\deg_X f(X,Y) = \deg_Y f(X,Y)$. If the tower $\mathcal{F}$ has finite genus $\gamma(\mathcal{F}) < \infty$, then it follows that*

$$V(\mathcal{F}) = V(\mathcal{G}) \quad \text{and} \quad V_w(\mathcal{F}) = V_w(\mathcal{G}).$$

Proof. We will prove here only the equality of the ramification loci $V(\mathcal{F}) = V(\mathcal{G})$; the assertion about the wild ramification loci is proved similarly (see [6]). We use all notations introduced above, cf. Figure 6.1. Suppose that $V(\mathcal{F}) \neq V(\mathcal{G})$; then we must show that the genus of the tower $\mathcal{F}$ satisfies $\gamma(\mathcal{F}) = \infty$. By symmetry, we can assume that there exists a place P of the function field $F_0 = G_0$ which is ramified in the tower $\mathcal{F}$, but it is unramified in the dual tower $\mathcal{G}$. Since the genera $g(F_n), g(G_n)$ and the extension degrees $[F_{n+1} : F_n]$ do not change under constant field extensions (see [48, p. 101 ff.]), we can replace the constant field $\mathbb{F}_q$ by its algebraic closure $\overline{\mathbb{F}}_q$ and we then assume that the fields F_n and G_n in the towers $\mathcal{F}$ and $\mathcal{G}$ are function fields over $\overline{\mathbb{F}}_q$.

The place P above is now a rational place of the function field F_0. We fix $k \geq 1$ and a place Q of F_k lying above P such that $e(Q|P) \geq 2$ (this exists since P is ramified in the tower $\mathcal{F}$). Let $m \geq 1$ and set $H_m := G_m \cdot F_k$. We have the following picture:

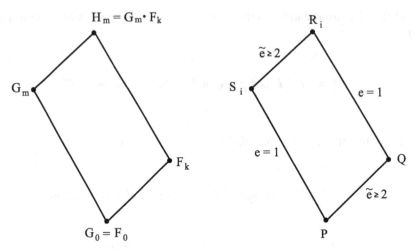

Figure 6.2

Setting $a := [F_1 : F_0] = [G_1 : F_0]$ (note that $\deg_X f = \deg_Y f$) we then have $[G_m : F_0] = [H_m : F_k] = a^m$ and $[F_k : F_0] = [H_m : G_m] = a^k$. The place Q of F_k splits completely in the extension H_m/F_k, since P splits completely in the extension G_m/F_0 (here we used that P is unramified in the tower $\mathcal{G}$, and that the constant field is algebraically closed). Denote by R_i the places of H_m above Q (for $i = 1, \ldots, a^m$), and by $S_i := R_i \cap G_m$ the restriction of R_i to the field G_m. Note that the places S_i for $i = 1, 2, \ldots, a^m$ are not necessarily distinct. Since S_i/P is unramified, it follows that the ramification index $\tilde{e}$ of $R_i|S_i$ is $\tilde{e} = e(R_i|S_i) = e(Q|P) \geq 2$ (see Figure 6.2). In particular, the different degree of the extension H_m/G_m satisfies

$$\deg \operatorname{Diff}(H_m/G_m) \geq \sum_{i=1}^{a^m} d(R_i|S_i) \geq a^m.$$

Observe now that we have field isomorphisms $H_m \simeq F_{m+k}$ and $G_m \simeq F_m$; together with the above inequality, this implies for all $m \geq 0$ the inequality

$$\deg \operatorname{Diff}(F_{m+k}/F_m) \geq a^m. \tag{6.4}$$

Now we consider the tower $\mathcal{E} = (E_0, E_1, E_2, \ldots)$ where $E_i := F_{k \cdot i}$ for all $i \geq 0$. It is clear that $\gamma(\mathcal{F}) = \gamma(\mathcal{E})$ and that $[E_{i+1} : E_i] = a^k$ for all $i \geq 0$. By (6.4) we have for all $n \geq 1$ the inequality

$$\deg \operatorname{Diff}(E_n/E_{n-1}) \geq a^{k(n-1)}.$$

Taking $\rho_n := a^{k(n-1)}$, it follows immediately from Theorem 6.2 that $\gamma(\mathcal{E}) = \infty$, and we then conclude that $\gamma(\mathcal{F}) = \infty$.   □

We illustrate the use of Theorem 6.6 with the following example.

Example 6.7. Let $q = p^3$ and consider the tower $\mathcal{F} = (F_0, F_1, F_2, \ldots)$ over $\mathbb{F}_q$ which is defined recursively by the equation

$$Y^{p+1} - Y + 1 = \frac{X^{p+1} - X + 1}{X^p}. \tag{6.5}$$

One can easily show that Equation (6.5) defines indeed a recursive tower. Let $\mathcal{G}$ denote the dual tower of $\mathcal{F}$. The following properties of the towers $\mathcal{F}$ and $\mathcal{G}$ are easily checked:

a) $V(\mathcal{F}) = V(\mathcal{G}) = \{(x_0 = 0), (x_0 = 1), (x_0 = \infty)\}$.

b) All places $(x_0 = \alpha)$ of F_0 with $\alpha^{p+1} - \alpha + 1 = 0$ split completely in the tower $\mathcal{F}$ (note here that all roots of $X^{p+1} - X + 1 = 0$ are in $\mathbb{F}_q$).

By Property b), the splitting rate $\nu(\mathcal{F})$ is at least $p + 1$, and by Property a) the ramification locus of the tower $\mathcal{F}$ is finite. So the tower $\mathcal{F}$ is a promising candidate for being asymptotically good over the cubic field $\mathbb{F}_q$ with $q = p^3$. However, the place $(x_0 = \infty)$ of $F_0 = G_0$ is tame in $\mathcal{F}$ and it is wild in the dual tower $\mathcal{G}$. Now it follows from Theorem 6.6 that the genus of $\mathcal{F}$ satisfies $\gamma(\mathcal{F}) = \infty$, and hence the tower $\mathcal{F}$ is asymptotically bad.

6.4 Classification of Recursive Artin-Schreier Towers

Let $p = \mathrm{char}(\mathbb{F}_q)$. A polynomial $\varphi(T) = T^{p^k} + a_{k-1}T^{p^{k-1}} + \ldots + a_0 T \in \mathbb{F}_q[T]$ is called an additive polynomial; it is separable if and only if $a_0 \neq 0$. We say that a tower $\mathcal{F} = (F_0, F_1, F_2, \ldots)$ over $\mathbb{F}_q$ is a recursive Artin-Schreier tower of degree p^k if $\mathcal{F}$ is defined recursively by an equation $\varphi(Y) = \psi(X)$, with $\varphi(Y) \in \mathbb{F}_q[Y]$ a separable monic additive polynomial of degree p^k and with $\psi(X) \in \mathbb{F}_q(X)$. Assuming that all roots of the polynomial $\varphi(T)$ belong to $\mathbb{F}_q$, then all steps F_n/F_{n-1} of the above tower are Galois extensions of degree p^k, each step having an elementary abelian Galois group of exponent p. Note that some of the asymptotically good wild towers which were presented in Section 5 are recursive Artin-Schreier towers.

Here we restrict our attention to recursive Artin-Schreier towers $\mathcal{F}$ over $\mathbb{F}_q$ of prime degree p. So $\mathcal{F}$ is defined recursively by an equation of the form

$$\varphi(Y) = Y^p + aY = \psi(X), \tag{6.6}$$

with $0 \neq a \in \mathbb{F}_q$ and with a rational function $\psi(X) \in \mathbb{F}_q(X)$. The following theorem shows that if $\mathcal{F}$ is asymptotically good, then the rational function $\psi(X)$ must be of a very special form.

Theorem 6.8. *Suppose that $\mathcal{F}$ is an asymptotically good Artin-Schreier tower over $\mathbb{F}_q$ which is recursively defined by Equation (6.6). Then the rational function $\psi(X)$ is of one of the following three types:*

i) $\psi(X) = (X-b)^p/\psi_1(X)+c$, *with elements $b, c \in \mathbb{F}_q$ and with a polynomial $\psi_1(X) \in \mathbb{F}_q[X] \setminus \mathbb{F}_q[X^p]$ of degree $\deg \psi_1(X) \leq p$ satisfying $\psi_1(b) \neq 0$.*

ii) $\psi(X) = \psi_0(X)/(X - b)^p$, *with an element $b \in \mathbb{F}_q$ and with a polynomial $\psi_0(X) \in \mathbb{F}_q[X] \setminus \mathbb{F}_q[X^p]$ of degree $\deg \psi_0(X) \leq p$ satisfying $\psi_0(b) \neq 0$.*

iii) $\psi(X) = 1/\psi_1(X) + c$, *with an element $c \in \mathbb{F}_q$ and with a polynomial $\psi_1(X) \in \mathbb{F}_q[X] \setminus \mathbb{F}_q[X^p]$ of degree $\deg \psi_1(X) = p$.*

For the proof of Theorem 6.8 we refer to [4]. Note that the examples in Section 5 of asymptotically good recursive Artin-Schreier towers are all of Type i). We do not know any example of an asymptotically good tower of Type ii) or Type iii).

6.5 h-Towers

We finish this review on towers of algebraic function fields over finite fields with another observation about recursive towers: many of the asymptotically good recursive towers known in the literature are so-called h-towers. This means that there exist a rational function $h(T)$ and two fractional linear transformations A and B in $\mathbb{F}_q(T)$ such that the tower is defined recursively by the equation

$$h(Y) = Ah(BX). \tag{6.7}$$

For instance, the towers $\mathcal{T}_1, \mathcal{T}_2$ and $\mathcal{T}_3$ in Section 3 and the towers $\mathcal{W}_1, \mathcal{W}_3, \mathcal{W}_4$ and $\mathcal{W}_5$ in Section 5 are h-towers for an appropriate choice of functions $h(T) \in \mathbb{F}_q(T)$.

In [7] some conditions for a tower to be an h-tower are investigated, especially for towers of Kummer or Artin-Schreier type. The transformation of a recursive tower (whenever possible) to a form as in Equation (6.7) is often desirable since it may facilitate both the consideration of the genus and of the splitting rate of the tower.

References

[1] P. Beelen, "Graphs and recursively defined towers of function fields", J. Number Theory, Vol. 108 , 217-240 (2004).

[2] S. Ballet, "Curves with many points and multiplication complexity in any extension of $\mathbb{F}_q$", Finite Fields Appl., Vol. 5, 364-377 (1999).

[3] P. Beelen and I. Bouw, "Asymptotically good towers and differential equations", Compositio Mathematica, Vol. 141, 1405-1424 (2005).

[4] P. Beelen, A. Garcia and H. Stichtenoth, "On towers of function fields of Artin-Schreier type", Bull. Braz. Math. Soc., Vol. 35, 151-164 (2004).

[5] P. Beelen, A. Garcia and H. Stichtenoth, "On towers of function fields over finite fields", *Arithmetic, Geometry and Coding Theory (AGCT 2003) (eds. Y. Aubry, G. Lachaud)*, Séminaires et Congrès 11, Société Mathématique de France, 1-20 (2005).

[6] P. Beelen, A. Garcia and H. Stichtenoth, "On ramification and genus of recursive towers", Portugal. Math., Vol. 62, 231-243 (2005).

[7] P. Beelen, A. Garcia and H. Stichtenoth, "Towards a classification of recursive towers of function fields over finite fields", Finite Fields Appl., Vol. 12, 56-77 (2006).

[8] J. Bezerra, A. Garcia, H. Stichtenoth, "An explicit tower of function fields over cubic finite fields and Zink's lower bound", J. Reine Angew. Math., Vol. 589, 159-199 (2005).

[9] J. Bezerra and A. Garcia, "A tower with non-Galois steps which attains the Drinfeld-Vladut bound", J. Number Theory, Vol. 106, 142-154 (2004).

[10] E. Bombieri, "Counting points on curves over finite fields", Sém. Bourbaki, Exp. No. 430, (1972/73).

[11] M. Deuring, "Die Typen der Multiplikatorenringe elliptischer Funktionenkörper", Abh. Math. Sem. Hamburg, Vol. 14, 197 - 272 (1941).

[12] V. G. Drinfeld and S. G. Vladut, "The number of points of an algebraic curve", Funktsional. Anal. i Prilozhen, Vol. 17, 68-69 (1983). [Funct. Anal. Appl., Vol. 17, 53-54 (1983).]

[13] I. Duursma, B. Poonen and M. Zieve, "Everywhere ramified towers of global function fields", *Proceedings of the 7th International Conference on Finite Fields and Applications Fq7, Toulouse (eds. G. Mullen et al.)*, LNCS, Vol. 2948, 148-153 (2004).

[14] N. Elkies, "Explicit modular towers", *Proceedings of the 35th Annual Allerton Conference on Communication, Control and Computing (eds. T. Basar et al.) Urbana, IL*, 23-32 (1997).

[15] N. Elkies, "Excellent nonlinear codes from modular curves", *STOC' 01: Proceedings of the 33rd Annual ACM Symp. on Theory of Computing, Hersonissos, Crete, Greece*, 200-208.

[16] N. Elkies, "Explicit towers of Drinfeld modular curves", *European Congress of Mathematics, vol.II (eds. C.Casacuberta et al.)*, Birkhäuser, Basel, 189-198 (2001).

[17] N. Elkies, "Still better nonlinear codes from modular curves", preprint, 2003.

[18] N. Elkies, E. Howe, A. Kresch, B. Poonen, J. Wetherell and M. Zieve, "Curves of every genus with many points, II: Asymptotically good families", Duke Mat. J., Vol. 122, 399-422 (2004).

[19] G. Frey, M. Perret and H. Stichtenoth, "On the different of abelian extensions of global fields", *Coding Theory and Algebraic Geometry. Proceedings, Luminy, 1991 (eds. H. Stichtenoth et al.)*, Springer LNM, Vol. 1518, 26 - 32 (1992).

[20] A. Garcia and H. Stichtenoth, "A tower of Artin-Schreier extensions of function fields attaining the Drinfeld-Vladut bound", Invent. Math., Vol. 121, 211-222 (1995).

[21] A. Garcia and H. Stichtenoth, "On the asymptotic behaviour of some towers of function fields over finite fields", J. Number Theory, Vol. 61, 248-273 (1996).

[22] A. Garcia, H. Stichtenoth, "Asymptotically good towers of function fields over finite fields", C. R. Acad. Sci. Paris, t. 322, Sèr. I, 1067-1070 (1996).

[23] A. Garcia, H. Stichtenoth, "Skew pyramids of function fields are asymptotically bad", *Coding Theory, Cryptography and Related Topics, Proceedings of a Conference in Guanajuato, 1998 (eds. J. Buchmann et al.)*, Springer-Verlag, Berlin, 111 - 113 (2000).

[24] A. Garcia and H. Stichtenoth, "On tame towers over finite fields", J. Reine Angew. Math., Vol. 557, 53-80 (2003).

[25] A. Garcia and H. Stichtenoth, "Asymptotics for the genus and the number of rational places in towers of function fields over a finite field", Finite Fields Appl., Vol. 11, 434-450 (2005).

[26] A.Garcia and H. Stichtenoth, "Some Artin-Schreier towers are easy", Moscow Math. J., to appear.

[27] A. Garcia and H. Stichtenoth, "On the Galois closure of towers", preprint, 2005.

[28] A. Garcia, H. Stichtenoth and M. Thomas, "On towers and composita of towers of function fields over finite fields", Finite Fields Appl., Vol. 3, 257-274 (1997).

[29] G. van der Geer, "Curves over finite fields and codes", *European Congress of Mathematics, vol.II (eds. C.Casacuberta et al.)*, Birkhäuser, Basel, 189-198 (2001).

[30] G. van der Geer and M. van der Vlugt, "An asymptotically good tower of curves over the field with eight elements", Bull. London Math. Soc., Vol. 34, 291-300 (2002).

[31] H. Hasse, "Zur Theorie der abstrakten elliptischen Funktionenkörper III", J. Reine Angew. Math., Vol. 175, 193 - 208 (1936).

[32] Y. Ihara, "Some remarks on the number of rational points of algebraic curves over finite fields", J. Fac. Sci. Tokyo, Vol. 28, 721 - 724 (1981).

[33] W.-C. W. Li, H. Maharaj and H. Stichtenoth, "New optimal towers of function fields over small finite fields", with an appendix by N.D. Elkies. *Proceedings of the 5th Int. Symp. on Algorithmic Number Theory, ANTS-5 (eds. C. Fieker and D. R. Kohel)*, LNCS, Vol. 2369, 372-389 (2002).

[34] S. Ling, H. Stichtenoth and S. Yang, "A class of Artin-Schreier towers with finite genus", Bull. Braz. Math. Soc., Vol. 36, 393-401 (2005).

[35] J. H. van Lint, *Introduction to Coding Theory*, Springer-Verlag, New York, 1982.

[36] H. W. Lenstra Jr., "On a problem of Garcia, Stichtenoth and Thomas", Finite Fields Appl., Vol. 8, 1-5 (2001).

[37] H. Maharaj and J. Wulftange, "On the construction of tame towers over finite fields", J. Pure Appl. Alg., Vol. 199, 197-218 (2005).

[38] Y. J. Manin, "What is the maximal number of points on a curve over $\mathbb{F}_2$?", J. Fac. Sci. Univ. Tokyo, Vol. 28, 715-720 (1981).

[39] R. Matsumoto, "Improvement of the Ashikmin-Litsyn-Tsfasman bound for quantum codes", IEEE Trans. Inform. Theory, Vol. 48, 2122-2124 (2002).

[40] H. Niederreiter and F. Özbudak, "Constructive asymptotic codes with an improvement on the Tsfasman-Vladut-Zink and Xing bound", *Coding, Cryptography and Combinatorics (eds. K. Q. Feng et al.)*, Progress in Computer Science and Applied Logic, Vol. 23, Birkhäuser, Basel, 259-275 (2004).

[41] H. Niederreiter and F. Özbudak, "Further improvements on asymptotic bounds for codes using distinguished divisors", Finite Fields Appl., to appear.

[42] H. Niederreiter and C. P. Xing, "Curve sequences with asymptotically many rational points", Contemporary Mathematics, Vol. 245, 3 - 14 (1999).

[43] H. Niederreiter and C. P. Xing, *Rational points of curves over finite fields*, Cambridge Univ. Press, Cambridge, 2001.

[44] H. Niederreiter and C. P. Xing, "Quasirandom points and global function fields", *Finite Fields and Applications*, Cambridge University Press, Cambridge, London Math. Soc. Lecture Notes Series, Vol. 233, 269-296 (1996).

[45] R. Pellikaan, H. Stichtenoth and F. Torres, "Weierstrass semigroups in an asymptotically good tower of function fields", Finite Fields Appl., Vol. 4, 381-392 (1998).

[46] J.-P. Serre, "Sur le nombre des points rationelles d'une courbe algébrique sur une corps fini", C. R. Acad. Sci. Paris, Vol. 296, 397 - 402 (1983).

[47] J.-P. Serre, "Rational points on curves over finite fields", unpublished lecture notes by F.Q. Gouvea, Harvard University, 1985.

[48] H. Stichtenoth, *Algebraic function fields and codes*, Springer Verlag, Berlin 1993.

[49] H. Stichtenoth, "Explicit constructions of towers of function fields with many rational places", *European Congress of Mathematics, Vol. II (eds. C.Casacuberta et al.)*, Birkhäuser, Basel, 219-224 (2001).

[50] H. Stichtenoth, "Transitive and self-dual codes attaining the Tsfasman-Vladut-Zink bound", IEEE Trans. Inform. Theory, Vol. 52, 2218-2214 (2006).

[51] H. Stichtenoth and C. P. Xing, "Excellent non-linear codes from algebraic function fields", IEEE Trans. Inform. Theory, Vol. 51, 4044-4046 (2005).

[52] K. O. Stöhr and J. F. Voloch, "Weierstrass points and curves over finite fields", Proc. London Math. Soc., Vol. 52, 1 - 19 (1986).

[53] A. Temkine, "Hilbert class field towers of function fields over finite fields and lower bounds on $A(q)$", J. Number Theory, Vol. 87, 189-210 (2001).

[54] M. A. Tsfasman, S. G. Vladut and T. Zink, "Modular curves, Shimura curves and Goppa codes better than the Varshamov-Gilbert bound", Math. Nachr., Vol. 109, 21-28 (1982).

[55] M. A. Tsfasman and S. G. Vladut, *Algebraic-geometric codes*, Kluwer Acad. Publ., Dordrecht/Boston/London, 1991.

[56] A. Weil, *Sur les courbes algébriques et les variétés qui s'en déduisent*, Act. Sc. et Industrielles, Vol. 1041. Hermann, Paris, 1948.

[57] J. Wulftange, *Zahme Türme algebraischer Funktionenkörper*, Ph.D. thesis, Essen, 2003.

[58] J. Wulftange, "On the construction of some towers over finite fields", *Proceedings of the 7th International Conference on Finite Fields and Applications Fq7, Toulouse, (eds. G.Mullen et al.)*, LNCS, Vol. 2948, 154-165 (2004).

[59] C. P. Xing, "Nonlinear codes from algebraic curves improving the Tsfasman-Vladut-Zink bound", IEEE Trans. on Information Theory, Vol. 49, 1653-1657 (2003).

[60] A. Zaytsev, "The Galois closure of the Garcia-Stichtenoth tower", preprint, 2005.

[61] T. Zink, "Degeneration of Shimura surfaces and a problem in coding theory", *Fundamentals of Computation Theory, (ed. L. Budach)*, LNCS, Vol. 199, 503-511(1985).

Chapter 2

FUNCTION FIELDS OVER FINITE FIELDS AND THEIR APPLICATIONS TO CRYPTOGRAPHY

Harald Niederreiter, Huaxiong Wang and Chaoping Xing

1. Introduction

It is well known that algebraic function fields over finite fields have many applications in coding theory, and the latter is closely related to cryptography. This has led researchers in a natural way to consider methods based on some specified function fields in order to construct cryptographic schemes, such as schemes for unconditionally secure authentication, traitor tracing, secret sharing, broadcast encryption and secure multicast, just to mention a few.

There is no doubt that the sophisticated techniques of function fields and their cryptographic applications have become a new, promising research direction. Over the past few years, a lot of research in this new direction has been carried out and the results are fruitful.

This paper surveys some initial efforts in this new emerging direction. We describe several interesting links among function fields, cryptography and combinatorics. As examples we show that constructions based on function fields for authentication codes, frameproof codes, perfect hash families, cover-free families and sequences with high linear complexity outperform the previously existing results, thus yielding in turn, directly or indirectly, cryptographic schemes with better performance.

It should be noted that this paper is far from exhaustive, as it presents only a portion of results that appeared in the literature. We hope that the paper will nevertheless stimulate further research in this new and promising direction of applying function fields to cryptography.

A. Garcia and H. Stichtenoth (eds.), Topics in Geometry, Coding Theory and Cryptography, 59–104.
© 2007 *Springer.*

2. Applications to Combinatorial Cryptography

In this section, we will present several interesting applications of function fields over finite fields to the construction of unconditionally secure authentication codes, frameproof codes, perfect hash families and cover-free families.

2.1 Constructions of Authentication Codes

One fundamental goal in cryptography is to ensure integrity of sensitive data, which simply means providing assurance about the content and origin of the communicated and/or stored data. Data integrity is accomplished by means such as digital signature schemes and message authentication codes. In a *digital signature scheme* the signature is generated using the secret key of the signer, and the authenticity is verified by a public verification algorithm. The security of signature schemes relies on some assumed computational complexity of problems such as the discrete logarithm and factorisation problems. A *message authentication code* (or MAC), on the other hand, is a private-key based cryptosystem, requiring to share a secret key between a sender and a receiver ahead of the communication. A typical example of a MAC is constructed by using block ciphers (e.g., DES or AES) in the cipher block chaining (CBC) mode. The MACs based on block ciphers are generally much faster than digital signature schemes, but there is no known proof of security, not even one based on a plausible computational assumption. However, it is possible to construct MACs that can be proved secure, without any computational assumptions. Such MACs are usually called *unconditionally secure authentication codes*, or *authentication codes*, or simply *A-codes* for short.

Unconditionally secure A-codes were invented by Gilbert, MacWilliams and Sloane [27]. The general theory of unconditional authentication was developed by Simmons ([59, 60]) and has been extensively studied in the past 25 years.

In the model for unconditional authentication, there are three participants: a *transmitter*, a *receiver* and an *opponent*. The transmitter wants to communicate some information to the receiver using a public channel which is subject to active attack. That is, the opponent can create a forged message and insert it into the channel. To protect against this threat, the transmitter and the receiver share a secret key which is then used in an authentication code.

A *systematic* A-code (or A-code *without secrecy*) is a code where the *source state* (i.e. plaintext) is concatenated with an *authenticator* (or a *tag*) to obtain a *message* which is sent through the channel. Such a code is a triple $(S, \mathcal{E}, \mathcal{T})$ of nonempty finite sets together with an authentication mapping $f : S \times \mathcal{E} \to \mathcal{T}$. Here S is the set of source states, $\mathcal{E}$ is the set of keys and $\mathcal{T}$ is the set of authenticators. When the transmitter wants to send the information $s \in S$ using a key $e \in \mathcal{E}$, which is secretly shared with the receiver, he transmits the message $m = (s, t)$, where $s \in S$ and $t = f(s, e) \in \mathcal{T}$. When the receiver gets a message $m = (s, t)$, she checks the authenticity by verifying whether $t = f(s, e)$ or not, using the secret key $e \in \mathcal{E}$.

The security of an A-code is measured by an opponent's deception probability. Suppose an opponent sees a sequence of $i \geq 0$ authenticated messages, all of which are authenticated using the same key. Then the opponent creates a new forged message, which he hopes is accepted by the receiver as authentic. For simplicity we only consider two typical attacks in the unconditionally secure authentication model. Suppose the opponent has the ability to insert messages into the channel and/or to modify existing messages. When the opponent inserts a new message $m' = (s', t')$ into the channel, this is called *impersonation attack*. When the opponent sees a message $m = (s, t)$ and changes it to a message $m' = (s', t')$ where $s \neq s'$, this is called *substitution attack*. We say a message (s, t) is *valid* if there exists a key $e \in \mathcal{E}$ such that $t = f(s, e)$.

We assume that there is a probability distribution on the source states which is known to all the participants. Given the probability distribution on the source states, the receiver and the transmitter will choose a probability distribution for $\mathcal{E}$. Denote the probability of success for the opponent when trying impersonation attack and substitution attack by P_I and P_S, respectively, and $P(\cdot)$ and $P(\cdot|\cdot)$ specify probability and conditional probability distribution on the message space $\mathcal{M}$. Then we have

$$P_I = \max_{s,t} P(m = (s, t) \text{ valid }),$$

$$P_S = \max_{s,t} \max_{\substack{s' \neq s \\ t'}} P(m' = (s', t') \text{ valid } | m = (s, t) \text{ observed }).$$

If we further assume that the keys and the source states are uniformly distributed, then the deception probabilities can be expressed as

$$P_I = \max_{s,t} \frac{|\{e \in \mathcal{E} : t = f(s, e)\}|}{|\mathcal{E}|},$$

$$P_S = \max_{s,t} \max_{\substack{s' \neq s \\ t'}} \frac{|\{e \in \mathcal{E} : t = f(s, e),\ t' = f(s', e)\}|}{|\{e \in \mathcal{E} : t = f(s, e)\}|}.$$

For the sake of simplicity, we will always assume that the keys and the source states are uniformly distributed.

For an A-code $(\mathcal{S}, \mathcal{E}, \mathcal{T})$, where $\mathcal{S}$ is a set of k source states and $\mathcal{T}$ is a set of ℓ authenticators, it is known [63] that $P_I \geq 1/\ell$ and $P_S \geq 1/\ell$. Codes with $P_I = P_S = 1/\ell$ have been known to be equivalent to orthogonal arrays. That is, we have the following result.

Lemma 2.1 ([63]). *Suppose we have an A-code $(\mathcal{S}, \mathcal{E}, \mathcal{T})$ without secrecy for k source states and having ℓ authenticators in which $P_I = P_S = 1/\ell$. Then*

$|\mathcal{E}| \geq k(\ell - 1) + 1$ *and equality occurs if and only if there exists an orthogonal array* $OA(\ell, k, \lambda)$, *where* $\lambda = (k(\ell - 1) + 1)/\ell^2$.

The goal of authentication theory is to derive bounds for various parameters in A-codes and to construct A-codes with desired properties. For reviews of different bounds and constructions for A-codes, we refer to [31, 32, 63].

One of the most powerful tools to construct A-codes is universal hash families introduced by Carter and Wegman [12] in 1979. It is worth mentioning that universal hash families have also found numerous applications in computer science, such as in complexity theory, search algorithms and information retrieval, to mention a few. We refer to [65] for a good account of the development in this field.

Consider a hash family $\mathcal{H}$, which is a set of N functions such that $h : A \to B$ for each $h \in \mathcal{H}$, where $|A| = k$ and $|B| = \ell$. There will be no loss in generality in assuming $k \geq \ell$ and we call $\mathcal{H}$ an $(N; k, \ell)$ *hash family*. We first review the relevant definitions and results as follows.

Definition 2.2. An $(N; k, \ell)$ hash family $\mathcal{H}$ is called ϵ-*almost universal* (ϵ-AU for short) if for any two distinct elements $a_1, a_2 \in A$, there are at most ϵN functions $h \in \mathcal{H}$ such that $h(a_1) = h(a_2)$.

The following lemma, due to Bierbrauer, Johansson, Kabatianskii and Smeets [4], establishes the equivalence between ϵ-AU hash families and error-correcting codes.

Lemma 2.3. *If there exists a q-ary code with codeword length N, cardinality M, and minimum Hamming distance d, then there exists an ϵ-AU $(N; M, q)$ hash family, where $\epsilon = 1 - d/N$. Conversely, if there exists an ϵ-AU $(N; M, q)$ hash family, then there exists a code with parameters as above.*

Definition 2.4. An $(N; k, \ell)$ hash family $\mathcal{H}$ is called ϵ-*almost strongly universal* (ϵ-ASU for short) if:

1 for any element $a \in A$ and any element $b \in B$, there exist exactly N/ℓ functions $h \in \mathcal{H}$ such that $h(a) = b$;

2 for any two distinct elements $a_1, a_2 \in A$ and for any two (not necessarily distinct) elements $b_1, b_2 \in B$, there exist at most $\epsilon N/\ell$ functions $h \in \mathcal{H}$ such that $h(a_i) = b_i, i = 1, 2$.

There is a strong relationship between ϵ-ASU hash families and A-codes. Given an A-code $(\mathcal{S}, \mathcal{E}, \mathcal{T})$ with authentication mapping $f : \mathcal{S} \times \mathcal{E} \to \mathcal{T}$ and with $P_I = 1/|\mathcal{T}|$ and P_S, each key $e \in \mathcal{E}$ corresponds to a unique function h_e from $\mathcal{S}$ to $\mathcal{T}$ defined by $h_e(s) = f(s, e)$. It is straightforward to verify that $\mathcal{H} = \{h_e : e \in \mathcal{E}\}$ is an ϵ-ASU hash family from $\mathcal{S}$ to $\mathcal{T}$, where $\epsilon = P_S$. Conversely, given an ϵ-ASU $(N; k, \ell)$ hash family $\mathcal{H}$ from A to B, we can associate an A-code $(\mathcal{S}, \mathcal{E}, \mathcal{T})$, where $\mathcal{S} = A$, $\mathcal{T} = B$ and $|\mathcal{E}| = |\mathcal{H}|$, and each key $e \in \mathcal{E}$ corresponds to a unique hash function $h_e \in \mathcal{H}$ indexed by e. The authentication mapping $f : \mathcal{S} \times \mathcal{E} \to \mathcal{T}$ is defined by $f(s, e) = h_e(s)$. Then the resulting A-code has $P_I = 1/\ell$ and $P_S \le \epsilon$. In summary, we have the following result.

Lemma 2.5 ([4, 64]). *If there exists an A-code $(\mathcal{S}, \mathcal{E}, \mathcal{T})$ with $P_I = 1/|\mathcal{T}|$ and P_S, then there exists an ϵ-ASU $(N; k, \ell)$ hash family, where $\epsilon = P_S$, $N = |\mathcal{E}|, k = |\mathcal{S}|$ and $\ell = |\mathcal{T}|$. Conversely, if there exists an ϵ-ASU $(N; k, \ell)$ hash family, then there exists an A-code with the above parameters such that $P_I = 1/|\mathcal{T}|$ and $P_S \le \epsilon$.*

Earlier research on A-codes had generally been devoted to constructions which ensure that the opponent's deception probabilities are bounded by $1/\ell$. In terms of ϵ-ASU hash families, this amounts to $\epsilon = 1/\ell$. Such codes were shown to be equivalent to an orthogonal array, and from Lemma 2.1 we know that $|\mathcal{E}| \ge k(\ell - 1) + 1$. This means that for a fixed security (i.e. $1/\ell$), the key size increases linearly as a function of the size of possible source states — a similar situation as for the "one-time pad". Thus, for a large size of the set of source states, it requires many bits of keys to store and "secretly" exchange.

The significance of ϵ-ASU hash families in the construction of A-codes, as observed by Wegman and Carter [76], is that by not requiring the deception probability to be the theoretical minimum, that is $\epsilon > 1/\ell$, we can expect to reduce the key size significantly. As shown in [4, 65, 76], by allowing $P_S > P_I$ (i.e. $\epsilon > 1/\ell$), it is possible that the size of the set of source states grows exponentially in the key size. This observation is very important from the viewpoint of practice, as we may deal with scenarios where we are satisfied with deception probabilities slightly larger than $1/\ell$, but where we have a limitation on the key storage.

A very useful method of constructing an ϵ-ASU hash family is to compose an AU hash family and an ASU hash family with appropriate parameters. The following lemma is due to Stinson [64] and independently to Bierbrauer, Johansson, Kabatianskii and Smeets [4].

Lemma 2.6 (Composition). *Let $\mathcal{H}_1$ be an ϵ_1-AU hash family from A_1 to B_1 and let $\mathcal{H}_2$ be an ϵ_2-ASU hash family from B_1 to B_2. Then $\mathcal{H} = \{h_2 h_1 : h_1 \in \mathcal{H}_1, h_2 \in \mathcal{H}_2\}$ is an ϵ-ASU hash family from A_1 to B_2 with $\epsilon = \epsilon_1 + \epsilon_2$.*

It turns out that many constructions of ϵ-ASU hash families with large $|A|$ are based on the above composition principle. The constructions giving good performance use Reed-Solomon codes [4, 64], or more general geometric codes [3], as the ϵ-AU hash family in the above composition construction. Helleseth and Johansson [29] gave a direct construction of ϵ-ASU hash families without using the above composition principle, rather by using exponential sums. For large $|A|$, it results in ϵ-ASU hash families with better performance.

In [88], Xing, Wang and Lam gave another direct construction (without using composition) of ϵ-ASU hash families from function fields over finite fields. The construction results in new classes of A-codes with better performance than those previously known. In the following, we describe the construction given in [88].

Let $F/\mathbb{F}_q$ be a global function field of genus g with $N(F) \geq 1$ and let T be a nonempty set of rational places of F. Let D be a positive divisor of F with $T \cap \mathrm{Supp}(D) = \emptyset$. Choose a rational place R in T and put $G = D - R$. Then $\deg(G) = \deg(D) - 1$, $\mathcal{L}(G) \subseteq \mathcal{L}(D)$ and $\mathbb{F}_q \cap \mathcal{L}(G) = \{0\}$. Let each pair $(P, \alpha) \in T \times \mathbb{F}_q$ be associated with a map $h_{(P,\alpha)}$ from $\mathcal{L}(G)$ to $\mathbb{F}_q$ defined by $h_{(P,\alpha)}(f) = f(P) + \alpha$.

Lemma 2.7. *Let $\mathcal{H} = \{h_{(P,\alpha)} : (P, \alpha) \in T \times \mathbb{F}_q\}$. If $\deg(D) \geq 2g + 1$, then the cardinality of $\mathcal{H}$ is equal to $q|T|$.*

Proof. It is sufficient to prove that $\{h_{(P,\alpha)}\}_{(P,\alpha) \in T \times \mathbb{F}_q}$ are pairwise distinct. Assume that $h_{(P,\alpha)} = h_{(Q,\beta)}$ for (P, α) and (Q, β) in $T \times \mathbb{F}_q$, i.e.,

$$h_{(P,\alpha)}(f) = h_{(Q,\beta)}(f) \tag{2.1}$$

for all $f \in \mathcal{L}(G)$. In particular,

$$\alpha = h_{(P,\alpha)}(0) = h_{(Q,\beta)}(0) = \beta. \tag{2.2}$$

It follows from (2.1) and (2.2) that

$$f(P) = f(Q)$$

for all $f \in \mathcal{L}(G)$. This implies that

$$e(P) = e(Q) \tag{2.3}$$

for all $e \in \mathcal{L}(D)$ since $\mathcal{L}(D) = \mathbb{F}_q \oplus \mathcal{L}(G)$.

Suppose that P is different from Q. As $\deg(D - P) > \deg(D - P - Q) \geq 2g - 1$, we obtain by the Riemann-Roch theorem

$$\ell(D - P) = \deg(D) - g, \quad \ell(D - P - Q) = \deg(D) - g - 1,$$

where $\ell(G) = \dim(\mathcal{L}(G))$ for a divisor G.

By the above results on dimensions, we can choose a function u from the set $\mathcal{L}(D - P) \setminus \mathcal{L}(D - P - Q)$. Then it is clear that $u(P) = 0$ and $u(Q) \neq 0$. This contradicts (2.3). Hence $P = Q$. The proof is complete. $\square$

Theorem 2.8 ([88]). *Let $F/\mathbb{F}_q$ be a global function field of genus g with $N(F) \geq 1$ and let T be a nonempty set of rational places of F. Suppose* *that D is a positive divisor of F with $\deg(D) \geq 2g+1$ and $T \cap \mathrm{Supp}(D) = \emptyset$. Then there exists an ϵ-ASU $(N; k, q)$ hash family with*

$$N = q|T|, \quad k = q^{\deg(D)-g}, \quad \epsilon = \frac{\deg(D)}{|T|}.$$

Proof. Let $R \in T$ be a rational place of F and put $G = D - R$. Define

$$A := \mathcal{L}(G), \quad B := \mathbb{F}_q$$

and

$$\mathcal{H} := \{h_{(P,\alpha)} : (P, \alpha) \in T \times \mathbb{F}_q\}.$$

Then $|\mathcal{H}| = q|T|$ by Lemma 2.7. It is easy to verify that for any element $a \in A = \mathcal{L}(G)$ and $b \in B = \mathbb{F}_q$, there exist exactly $|T| = N/q$ pairs $(P, \alpha) \in T \times \mathbb{F}_q$ such that

$$h_{(P,\alpha)}(a) = a(P) + \alpha = b,$$

i.e., there exist exactly N/q functions $h_{(P,\alpha)} \in \mathcal{H}$ such that $h_{(P,\alpha)}(a) = b$.

Now let a_1, a_2 be two distinct elements of A and b_1, b_2 two elements of B. We consider

$$
\begin{aligned}
m \ := \ & \max_{\substack{a_1 \neq a_2 \in A \\ b_1, b_2 \in B}} \left| \left\{ h_{(P,\alpha)} \in \mathcal{H} \ : \ h_{(P,\alpha)}(a_1) = b_1; h_{(P,\alpha)}(a_2) = b_2 \right\} \right| \\
= \ & \max_{\substack{a_1 \neq a_2 \in A \\ b_1, b_2 \in B}} \left| \left\{ (P, \alpha) \in T \times \mathbb{F}_q \ : \ a_1(P) + \alpha = b_1; a_2(P) + \alpha = b_2 \right\} \right| \\
= \ & \max_{\substack{a_1 \neq a_2 \in A \\ b_1, b_2 \in B}} \left| \left\{ (P, \alpha) \in T \times \mathbb{F}_q \ : \ \begin{matrix} (a_1 - a_2 - b_1 + b_2)(P) = 0; \\ a_2(P) + \alpha = b_2 \end{matrix} \right\} \right|.
\end{aligned}
$$

As $a_1 - a_2 \in \mathcal{L}(G) \setminus \{0\}$ and $b_1 - b_2 \in \mathbb{F}_q$, we know that $0 \neq a_1 - a_2 - b_1 + b_2 \in \mathcal{L}(D)$. Thus, there are at most $\deg(D)$ distinct zeros of $a_1 - a_2 - b_1 + b_2$ in T. Since α is uniquely determined by P from the equality $a_2(P) + \alpha = b_2$, we have that at most $\deg(D)$ pairs $(P, \alpha) \in T \times \mathbb{F}_q$ satisfy

$$(a_1 - a_2 - b_1 + b_2)(P) = 0 \quad \text{and} \quad a_2(P) + \alpha = b_2,$$

i.e.,

$$m \leq \deg(D) = \frac{\deg(D)}{|T|} \cdot \frac{N}{q}.$$

Hence we can take $\epsilon = \deg(D)/|T|$. This completes the proof. $\qquad\square$

Example 2.9. Consider the rational function field $F/\mathbb{F}_q$. Then $g = g(F) = 0$.

(a) Let d be an integer between 1 and q and P a rational place of F. Put $D = dP$ and let T be the set of all rational places of F except P. Then

$\deg(D) = d \geq 2g + 1$, $|T| = q$ and $T \cap \mathrm{Supp}(D) = \emptyset$. By Theorem 2.8, we obtain an ϵ-ASU $(N; k, q)$ hash family with

$$N = q^2, \quad k = q^d, \quad \epsilon = \frac{d}{q}.$$

The ϵ-ASU hash family with the above parameters can also be found in [4].

(b) Let d be an integer between 2 and q and let T be the set of all rational places of F. As there always exists an irreducible polynomial of degree d over $\mathbb{F}_q$, we can find a positive divisor D such that $\deg(D) = d$ and $T \cap \mathrm{Supp}(D) = \emptyset$. Then $\deg(D) = d \geq 2g + 1$ and $|T| = q + 1$. By Theorem 2.8, we obtain an ϵ-ASU $(N; k, q)$ hash family with

$$N = q(q + 1), \quad k = q^d, \quad \epsilon = \frac{d}{q + 1}.$$

Example 2.10. Let the prime power q be a square and put $r = \sqrt{q}$. Consider a sequence of function fields $F_m/\mathbb{F}_q$ given in [24] as follows (see also Chapter 1, Section 5.1). Let F_1 be the rational function field $F_1 = \mathbb{F}_q(x_1)$. For $m \geq 2$ let $F_m = F_{m-1}(x_m)$ with

$$x_m^r + x_m = \frac{x_{m-1}^r}{x_{m-1}^{r-1} + 1}.$$

Then the number of rational places of F_m is more than $(r - 1)r^m$ and the genus g_m of F_m is less than r^m for all $m \geq 1$. Choose an integer c between 2 and $\sqrt{q} - 1$ (c is independent of m) and a rational place P_m of F_m and put $D_m = cq^{m/2}P_m$. Let T_m be a subset of $F_m(\mathbb{F}_q) \setminus \{P_m\}$ with

$$|T_m| = (r - 1)r^m = q^{m/2}(\sqrt{q} - 1),$$

where $F_m(\mathbb{F}_q)$ denotes the set of rational places of F_m. By Theorem 2.8, we obtain a sequence of ϵ-ASU $(N_m; k_m, q)$ hash families with

$$N_m = q^{m/2}(q\sqrt{q} - q), \quad k_m = q^{cq^{m/2}-g_m} > q^{(c-1)q^{m/2}}, \quad \epsilon = \frac{c}{\sqrt{q} - 1}.$$

By phrasing the construction in the above example in terms of A-codes, we obtain the following result.

Corollary 2.11. *The construction in Example 2.10 results in an A-code* $C = (S, \mathcal{E}, T)$ *with the parameters*

$$|S| = q^{(c-1)q^{m/2}}, \quad |\mathcal{E}| = q^{m/2}(q\sqrt{q} - q), \quad |T| = q,$$

and with deception probabilities

$$P_I = \frac{1}{q}, \; P_S \leq \frac{c}{\sqrt{q} - 1},$$

where q is a prime power and a square, whereas c and m are integers satisfying $2 \leq c \leq \sqrt{q} - 1$ and $m \geq 1$.

It has been shown in [88] that the above A-codes outperform some best known A-codes for several parameter sets.

2.2   Constructions of Frameproof Codes

In order to protect copyrighted material (such as digital data, computer software, etc.), a distributor endows each copy with some codeword that allows him/her to detect any unauthorised copy and trace it back to the user who created it. Such codes can offer protection by providing some form of traceability for pirated data. Frameproof codes are such codes first introduced by Boneh and Shaw [10] in the context of digital fingerprinting.

We begin with an example to motivate the topic. In the broadcast encryption scheme suggested by Chor, Fiat and Naor in [13], a sender (such as a distribution centre or a company) wishes to broadcast an encrypted message to a set of users who use their individual decoder to decrypt the broadcast message. A decoder box consists of N keys, where each key takes on one of q possible values. A decoder box a can typically be represented as an N-tuple $(a_1, \ldots, a_N)$, where $1 \leq a_i \leq q$ for $1 \leq i \leq N$. A user can redistribute the key of his decoder box without altering it. If an unauthorised copy of the decoder box is found containing the keys of user u's decoder box, we can accuse u of producing a *pirate* decoder box. However, u could claim that he was framed by a coalition which created a decoder box containing his keys. Thus, it is desirable to construct decoder boxes that satisfy the following property: *no coalition can collude to frame a user not in the coalition.* Codes that satisfy this property are then called *frameproof codes*, or *c-frameproof codes* if the condition is relaxed by limiting the size of the coalition to at most c users.

We will use a slightly different definition from the one in the literature, by using the restriction that the keys are chosen from a finite field.

Let $\mathbb{F}_q$ denote again the finite field with q elements and let N be a positive integer. Define the i-th projection

$$\pi_i : \mathbb{F}_q^N \rightarrow \mathbb{F}_q, \quad (a_1, \ldots, a_N) \mapsto a_i.$$

For a subset $A \subseteq \mathbb{F}_q^N$, we define the *descendants* of A, $\mathrm{desc}(A)$, to be the set of all $\mathbf{x} \in \mathbb{F}_q^N$ such that for each $1 \leq i \leq N$ there exists $\mathbf{a} \in A$ satisfying $\pi_i(\mathbf{x} - \mathbf{a}) = 0$.

Definition 2.12. Let $c \geq 2$ be an integer. A *q-ary c-frameproof code* of length N is a subset $C \subseteq \mathbb{F}_q^N$ such that for all $A \subseteq C$ with $|A| \leq c$ we have

$$\operatorname{desc}(A) \cap C = A.$$

From this definition it is clear that a q-ary c-frameproof code C is a q-ary c_1-frameproof code for any $2 \leq c_1 \leq c$.

We denote a q-ary c-frameproof code in $\mathbb{F}_q^N$ of size M by c-$FPC(N, M)$. Since each codeword can be "fingerprinted" into a copy of the distributed documents, we would like the value of M as large as possible. This leads to the following definition.

Definition 2.13. For a fixed prime power q and integers $c \geq 2$ and $N \geq 2$, let $M_q(N, c)$ denote the maximal size of q-ary c-frameproof codes of length N, i.e.,

$$M_q(N, c) := \max\{M : \text{there exists a } q\text{-ary } c\text{-}FPC(N, M)\}.$$

In [61] Staddon, Stinson and Wei proved that $M_q(N, c) \leq c\left(q^{\lceil N/c \rceil} - 1\right)$, which was improved by Blackburn in [7] where the following was shown.

Theorem 2.14. *Let* $c \geq 2$ *and* $N \geq 2$ *be integers and* q *a prime power. Then*

$$M_q(N, c) \leq \max\left\{q^{\lceil N/c \rceil}, r\left(q^{\lceil N/c \rceil} - 1\right) + (c - r)\left(q^{\lfloor N/c \rfloor} - 1\right)\right\},$$

where r *is the unique integer in* $\{0, 1, \ldots, c - 1\}$ *such that* $r \equiv N \pmod{c}$.

Consider the following construction in [7]. Let C be the set of all codewords of length N and weight exactly 1. Then C is a c-frameproof code of cardinality $N(q - 1)$. Indeed, let $a \in C$ with its ith component nonzero. Now, any set $A \subseteq C$ such that $a \in \operatorname{desc}(A)$ must contain a codeword b such that $a_i = b_i$. Since a codeword of weight 1 is uniquely determined by its nonzero component, it follows that $a = b$. Hence C is a c-frameproof code. Combining Theorem 2.14 and this construction, we obtain that for any integers $2 \leq N \leq c$ and for any prime power q we have $M_q(N, c) = N(q-1)$. Thus, we shall be interested only in the case where $N > c$.

We consider now the asymptotic behaviour of $M_q(N, c)$. From Theorem 2.14, we have $M_q(N, c) = \theta(q^{\lceil N/c \rceil})$, and the hidden constants in the approximation notation may depend on N, c and q. This suggests the ratio $D_q(c)$ defined as follows. For fixed q and c, define the asymptotic quantity

$$D_q(c) = \limsup_{N \to \infty} \frac{\log_q M_q(N, c)}{N}.$$

The following upper bound on $D_q(c)$ can be derived from [7].

Theorem 2.15. *We always have*

$$D_q(c) \leq \frac{1}{c}.$$

It has been suggested by many authors that frameproof codes can be constructed from error-correcting codes. We review the construction based on *linear* error-correcting codes proposed by Cohen and Encheva [14], as the constructions from function fields can be viewed as a generalisation of this construction.

Proposition 2.16. *Let q be a prime power. Then a q-ary $[N, k, d]$-linear code C is a q-ary c-$FPC(N, q^k)$ with $c = \lfloor (N-1)/(N-d) \rfloor$.*

Proof. Let A be a subset of C with $|A| \leq c$. Suppose that $\mathrm{desc}(A) \cap C \neq A$. Choose $\mathbf{x} \in (\mathrm{desc}(A) \cap C) \setminus A$. Since there are at most c elements in A, it follows from $\mathbf{x} \in \mathrm{desc}(A)$ that there is a codeword $\mathbf{y} \in A$ such that $\mathbf{y}$ agrees with $\mathbf{x}$ in at least $\lceil N/c \rceil$ positions, i.e., the Hamming weight of $\mathbf{y} - \mathbf{x}$ satisfies

$$wt(\mathbf{y} - \mathbf{x}) \leq N - \lceil \frac{N}{c} \rceil \leq N - \frac{N}{c}.$$

As $\mathbf{x} \neq \mathbf{y}$, we get

$$d \leq wt(\mathbf{y} - \mathbf{x}) \leq N - \frac{N}{c},$$

which implies $c \geq N/(N-d)$. This contradicts $c = \lfloor (N-1)/(N-d) \rfloor$. □

We remark that in this construction, as well as other constructions from error-correcting codes, the crucial parameter c is determined by the minimum distance of C if the length is given. Xing [82] showed that this feature can be relaxed based on global function fields, and consequently the results from error-correcting codes can be improved.

From the above relationship between linear codes and frameproof codes, we immediately obtain a lower bound on $D_q(c)$ using the Gilbert-Varshamov bound.

Theorem 2.17. *Let q be a prime power and $c \geq 2$ an integer. Then*

$$D_q(c) \geq 1 - H_q(1 - \frac{1}{c}),$$

where

$$H_q(\delta) = \delta \log_q(q-1) - \delta \log_q \delta - (1-\delta) \log_q(1-\delta)$$

is the q-ary entropy function.

Note that the bound in Theorem 2.17 is only an existence result as the Gilbert-Varshamov bound is not constructive.

Xing [82] gave two new lower bounds on $D_q(c)$ from algebraic-geometry codes. One bound can be obtained by directly applying Proposition 2.16 and the Tsfasman-Vlăduţ-Zink bound, while the other bound relies on the Jacobian group structure of global function fields.

From the theory of algebraic-geometry codes, we have the following lemma (see Chapter 1 and [46, Theorem 6.2.6]).

Lemma 2.18. *For any prime power q and any $0 \le \delta < 1$, there exists a sequence of q-ary (N_i, M_i, d_i)-codes such that $N_i \to \infty$ as $i \to \infty$ and*

$$\lim_{i \to \infty} \frac{d_i}{N_i} = \delta, \quad \lim_{i \to \infty} \frac{\log_q M_i}{N_i} \ge 1 - \delta - \frac{1}{A(q)}.$$

Here and in the following, $A(q)$ is the well-known quantity

$$A(q) := \limsup_{g \to \infty} \frac{1}{g} \max_{\substack{F/\mathbb{F}_q \\ g(F/\mathbb{F}_q)=g}} N(F/\mathbb{F}_q) \tag{2.4}$$

from the asymptotic theory of rational places of global function fields (see Chapter 1 and [46, Chapter 5]).

A direct consequence of the above lemma leads to an improved new lower bound on $D_q(c)$.

Theorem 2.19. *For any prime power q and any integer $c \ge 2$, we have*

$$D_q(c) \ge \frac{1}{c} - \frac{1}{A(q)}.$$

Note that (i) the bound in Theorem 2.19 is constructive since the algebraic-geometry codes in Lemma 2.18 are constructive as long as the sequences of global function fields are explicit (see Chapter 1, Sections 4 and 5); (ii) comparing with the upper bound in Theorem 2.15, we find that

$$\frac{1}{c} - \frac{1}{A(q)} \le D_q(c) \le \frac{1}{c}.$$

As we can see from the following corollary, $1/A(q) \to 0$ as $q \to \infty$, i.e., $D_q(c)$ is getting closer to $1/c$ as $q \to \infty$.

Corollary 2.20. (i) *If q is a prime power square, then for any $c \ge 2$,*

$$D_q(c) \ge \frac{1}{c} - \frac{1}{\sqrt{q} - 1}.$$

(ii) *If q is a prime power cube, then for any c ≥ 2,*

$$D_q(c) \geq \frac{1}{c} - \frac{q^{1/3} + 2}{2(q^{2/3} - 1)}.$$

(iii) *If q is any prime power, then for any c ≥ 2,*

$$D_q(c) \geq \frac{1}{c} - \frac{96}{\log_2 q}.$$

Another improved lower bound on $D_q(c)$ is obtained by applying algebraic-geometry codes based on the Jacobian group structure of global function fields.

Let $P_1, P_2, \ldots, P_N$ be N distinct rational places of $F/\mathbb{F}_q$. Choose a positive divisor D of F such that $\mathcal{L}(D - \sum_{i=1}^{N} P_i) = \{0\}$. Let $\nu_{P_i}(D) = v_i \geq 0$ and t_i be a local parameter at P_i for each $i = 1, 2, \ldots, N$.

Consider the map

$$\phi : \mathcal{L}(D) \longrightarrow \mathbb{F}_q^N, \quad f \mapsto ((t_1^{v_1} f)(P_1), (t_2^{v_2} f)(P_2), \ldots, (t_N^{v_N} f)(P_N)).$$

Then the image of ϕ forms a subspace of $\mathbb{F}_q^N$ that is defined as an algebraic-geometry code. The image of ϕ is denoted by $C(P_1, P_2, \ldots, P_N; D)_L$, or simply $C(\sum_{i=1}^{N} P_i, D)_L$. The map ϕ is an embedding since $\mathcal{L}(D - \sum_{i=1}^{N} P_i) = \{0\}$. Thus, the dimension of $C(P_1, P_2, \ldots, P_N; D)_L$ is equal to $\ell(D) := \dim(\mathcal{L}(D))$.

Note that the above construction is a modified version of algebraic-geometry codes defined by Goppa. The advantage of this construction is to make it possible to get rid of the condition $\text{Supp}(D) \cap \{P_1, P_2, \ldots, P_N\} = \emptyset$. This is crucial for this construction of frameproof codes. When the condition $\text{Supp}(D) \cap \{P_1, P_2, \ldots, P_N\} = \emptyset$ is satisfied, i.e., $v_i = 0$ for all $i = 1, 2, \ldots, N$, then the above construction of algebraic-geometry codes agrees with Goppa's construction.

Proposition 2.21. *Let $F/\mathbb{F}_q$ be a function field and let $P_1, P_2, \ldots, P_N$ be N distinct rational places of F. Let D be a positive divisor of F such that $\deg(D) < N$. Let $c \geq 2$ satisfy $\mathcal{L}(cD - \sum_{i=1}^{N} P_i) = \{0\}$. Then $C(\sum_{i=1}^{N} P_i, D)_L$ is a c-FPC$(N, q^{\ell(D)})$.*

Proof. Denote by $\mathbf{c}_f$ the codeword

$$\phi(f) = ((t_1^{v_1} f)(P_1), (t_2^{v_2} f)(P_2), \ldots, (t_N^{v_N} f)(P_N)) \quad \text{for all } f \in \mathcal{L}(D).$$

Let $A = \{\mathbf{c}_{f_1}, \ldots, \mathbf{c}_{f_r}\}$ be a subset of $C := C(P_1, P_2, \ldots, P_N; D)_L$ with $|A| = r \leq c$. Let $\mathbf{c}_h \in \text{desc}(A) \cap C$ for some $h \in \mathcal{L}(D)$. Then by the definition of descendant, for each $1 \leq i \leq N$ we have

$$\prod_{j=1}^{r} \pi_i(\mathbf{c}_{f_j} - \mathbf{c}_h) = 0,$$

where $\pi_i(\mathbf{c}_{f_j} - \mathbf{c}_h)$ stands for ith coordinate of $\mathbf{c}_{f_j} - \mathbf{c}_h$. This implies that

$$\prod_{j=1}^{r}(t_i^{v_i}f_j - t_i^{v_i}h)(P_i) = 0,$$

i.e.,

$$\nu_{P_i}(\prod_{j=1}^{r}(t_i^{v_i}f_j - t_i^{v_i}h)) \geq 1.$$

This is equivalent to

$$\nu_{P_i}(\prod_{j=1}^{r}(f_j - h)) \geq -rv_i + 1.$$

Hence

$$\prod_{j=1}^{r}(f_j - h) \in \mathcal{L}(rD - \sum_{i=1}^{N}P_i) \subseteq \mathcal{L}(cD - \sum_{i=1}^{N}P_i) = \{0\}.$$

Thus, the function $\prod_{j=1}^{r}(f_j - h)$ is the zero function. So, $f_j - h = 0$ for some $1 \leq j \leq r$. Hence $\mathbf{c}_h = \mathbf{c}_{f_j} \in A$. $\qquad\square$

We remark that in the above proposition the minimum distance of the algebraic-geometry code does not play an important role, in contrast to the general construction based on error-correcting codes. However, the result in Proposition 2.21 is not constructive.

From Proposition 2.21, we see that it is crucial to find a positive divisor D such that $\mathcal{L}(cD - \sum_{i=1}^{n}P_i) = \{0\}$. We will show some sufficient conditions for the existence of such divisors D. But first let us introduce some notation.

For a function field $F/\mathbb{F}_q$ of genus g, let $A_i(F)$ (or simply denoted by A_i in the case of no confusion) be the number of all positive divisors of F of fixed degree $i \geq 0$. It is clear that $A_0 = 1$ and A_1 is the number of rational places of $F/\mathbb{F}_q$. The zeta-function of $F/\mathbb{F}_q$ is defined as the power series

$$Z(T) := \sum_{i=0}^{\infty}A_iT^i.$$

The zeta-function can be written as a rational function

$$Z(T) = \frac{L(T)}{(1 - T)(1 - qT)},$$

where $L(T)$ is a polynomial of degree $2g$ with integral coefficients. All roots of $T^{2g}L(1/T)$ have absolute value $\sqrt{q}$. The polynomial $L(T)$ is called the *L-polynomial* of $F/\mathbb{F}_q$.

Let $J(F/\mathbb{F}_q)$ denote the zero divisor class group of $F/\mathbb{F}_q$. It is a finite abelian group. The order of $J(F/\mathbb{F}_q)$ is $L(1)$, where $L(T)$ is the L-polynomial of $F/\mathbb{F}_q$. Suppose that $F/\mathbb{F}_q$ has a rational place P_0. For a divisor D of F, we denote by $[D - \deg(D)P_0]$ the class of the zero divisor $D - \deg(D)P_0$ in $J(F/\mathbb{F}_q)$.

Lemma 2.22. *Let $F/\mathbb{F}_q$ be a function field of genus g with a rational place P_0. Then for an integer $c \geq 2$ and any fixed integer $m \geq g$, the subgroup of $J(F/\mathbb{F}_q)$ given by*

$$\{c[D - \deg(D)P_0] : D \geq 0, \deg(D) = m\}$$

has order at least h/c^{2g}, where h denotes the zero divisor class number $|J(F/\mathbb{F}_q)| = L(1)$.

Proof. It is a well-known fact that the set

$$\{[D - \deg(D)P_0] : D \geq 0, \deg(D) = m\}$$

is the whole group $J(F/\mathbb{F}_q)$ for $m \geq g$ (e.g. see [80, proof of Lemma 2.2]). Therefore,

$$\{c[D - \deg(D)P_0] : D \geq 0, \deg(D) = m\} = cJ(F/\mathbb{F}_q).$$

Since the p-rank of $J(F/\mathbb{F}_q)$ is at most $2g$ for any prime p (see [41, p. 39]), the desired result follows. □

Lemma 2.23. *Let $F/\mathbb{F}_q$ be a function field of genus g with at least one rational place P_0. Let c, m, n be three integers satisfying $c \geq 2$ and $g \leq m \leq n < cm$ and G a fixed positive divisor of F of degree n. Then there exists a positive divisor D of F of degree m such that $\mathcal{L}(cD - G) = \{0\}$ provided that $A_{cm-n} < h/c^{2g}$.*

Proof. By Lemma 2.22, we have

$$|\{[cH - c\deg(H)P_0] : H \geq 0, \deg(H) = m\}| = |cJ(F/\mathbb{F}_q)| \geq \frac{h}{c^{2g}}.$$

Moreover,

$$|\{[K + G - \deg(K + G)P_0] : K \geq 0, \deg(K) = cm - n\}| \leq A_{cm-n}.$$

Thus, $\{[cH - c\deg(H)P_0] : H \geq 0, \deg(H) = m\} \setminus \{[K + G - \deg(K + G)P_0] : K \geq 0, \deg(K) = cm - n\}$ is not empty. Choose an element $[cD - c\deg(D)P_0]$ from the above nonempty set for some positive divisor D of F. We claim that $\mathcal{L}(cD - G) = \{0\}$. Otherwise, there would be a nonzero function $f \in \mathcal{L}(cD - G)$. Therefore, the divisor $\operatorname{div}(f) + cD - G$ is a positive divisor.

Put $K = \text{div}(f) + cD - G$. Then $\deg(K) = cm - n$ and cD is equivalent to $K + G$, i.e., $[cD - c\deg(D)P_0]$ is the same as $[K + G - \deg(K+G)P_0]$. This contradicts the choice of $[cD - c\deg(D)P_0]$. $\qquad\square$

Lemma 2.24. *Let $F/\mathbb{F}_q$ be a function field of genus $g \geq 2$ with zero divisor class number h. Then the number A_r of positive divisors of $F/\mathbb{F}_q$ of degree r satisfies*

$$A_r < \frac{(3\sqrt{q} - 1)q^{r+1-g}gh}{(q-1)(\sqrt{q}-1)} \qquad \text{for } 0 \leq r \leq g - 1.$$

Proof. It was shown in [44, Lemma 3(ii)] that in the field $\mathbb{C}(z)$ of rational functions over the complex numbers we have the identity

$$\sum_{n=0}^{g-2} A_n z^n + \sum_{n=0}^{g-1} q^{g-1-n} A_n z^{2g-2-n} = \frac{L(z) - hz^g}{(1-z)(1-qz)},$$

where $L(z)$ is the L-polynomial of $F/\mathbb{F}_q$. Letting $z \to 1$ we get

$$\begin{aligned}
\sum_{n=0}^{g-2} A_n + \sum_{n=0}^{g-1} q^{g-1-n} A_n &= \lim_{z \to 1} \frac{L(z) - hz^g}{(1-z)(1-qz)} \\
&= \lim_{z \to 1} \frac{L'(z) - ghz^{g-1}}{-(1-qz) - q(1-z)} \\
&= \frac{L'(1) - gh}{q - 1}.
\end{aligned}$$

Now $L(z) = \prod_{i=1}^{2g}(1 - \omega_i z)$ with $|\omega_i| = \sqrt{q}$ for $1 \leq i \leq 2g$ by Weil's proof of the Riemann hypothesis for global function fields. By logarithmic differentiation,

$$\frac{L'(z)}{L(z)} = \sum_{i=1}^{2g} \frac{-\omega_i}{1 - \omega_i z}.$$

Putting $z = 1$ and using $L(1) = h$, we obtain

$$|L'(1)| \leq h \sum_{i=1}^{2g} \frac{|\omega_i|}{|1 - \omega_i|} \leq \frac{2gh\sqrt{q}}{\sqrt{q} - 1},$$

and so

$$\left| \frac{L'(1) - gh}{q-1} \right| \leq \frac{(3\sqrt{q} - 1)gh}{(q-1)(\sqrt{q}-1)}.$$

Thus, by noting that

$$q^{g-1-r} A_r < \sum_{n=0}^{g-2} A_n + \sum_{n=0}^{g-1} q^{g-1-n} A_n \leq \frac{(3\sqrt{q} - 1)gh}{(q-1)(\sqrt{q}-1)}$$

for $0 \leq r \leq g - 1$, we complete the proof.   □

Lemma 2.25. *Let $F/\mathbb{F}_q$ be a function field of genus $g \geq q$ with at least one rational place. Let c, m, n be three integers satisfying $c \geq 2$ and $g \leq m \leq n < cm$ and*

$$cm - n \leq g(1 - 2\log_q c) - 1 - \log_q \frac{(3\sqrt{q} - 1)g}{(q - 1)(\sqrt{q} - 1)}. \tag{2.5}$$

Let G be a fixed positive divisor of F of degree n. Then there exists a positive divisor D of F of degree m such that $\mathcal{L}(cD - G) = \{0\}$.

Proof. By rewriting the inequality (2.5), we have

$$\frac{(3\sqrt{q} - 1)gq^{(cm-n)+1-g}h}{(q - 1)(\sqrt{q} - 1)} \leq \frac{h}{c^{2g}}.$$

The desired result follows now from Lemmas 2.23 and 2.24.   □

Theorem 2.26. *For any prime power q and any integer $2 \leq c < \sqrt{q}$, we have*

$$D_q(c) \geq \frac{1}{c} - \frac{1}{A(q)} + \frac{(1 - 2\log_q c)}{c} \cdot \frac{1}{A(q)}.$$

Proof. Choose a family of function fields $F_i/\mathbb{F}_q$ with growing genus such that $\lim_{i \to \infty} N(F_i)/g(F_i) = A(q)$. Put $n_i = N(F_i)$ and $g_i = g(F_i)$. Let $\mathcal{P}_i$ be the set of all rational places of F_i and put

$$G_i = \sum_{P \in \mathcal{P}_i} P.$$

For any fixed $\varepsilon > 0$, put

$$m_i = \left\lfloor \frac{n_i}{c} + \frac{(1 - 2\log_q c)g_i}{c} - \frac{\varepsilon g_i}{c} \right\rfloor.$$

Then

$$\lim_{i \to \infty} \frac{cm_i - n_i - (1 - 2\log_q c)g_i}{g_i} = -\varepsilon < 0.$$

Therefore, for all sufficiently large i we have

$$cm_i - n_i \leq g_i(1 - 2\log_q c) - 1 - \log_q \frac{(3\sqrt{q} - 1)g_i}{(q - 1)(\sqrt{q} - 1)}.$$

By Lemma 2.25, there exists a positive divisor D_i of F_i of degree m_i such that $\mathcal{L}(cD_i - G_i) = \{0\}$ for each sufficiently large i. Thus, by Proposition 2.21 the code $C(G_i, D_i)_L$ is a c-$FPC(n_i, q^{\ell(D_i)})$. Hence

$$
\begin{aligned}
D_q(c) &\geq \limsup_{i \to \infty} \frac{\log_q q^{\ell(D_i)}}{n_i} \\
&\geq \lim_{i \to \infty} \frac{m_i - g_i + 1}{n_i} \\
&= \frac{1}{c} - \frac{1}{A(q)} + \frac{(1 - 2\log_q c)}{c} \cdot \frac{1}{A(q)} - \frac{\varepsilon}{cA(q)}.
\end{aligned}
$$

Since the above inequality holds for any $\varepsilon > 0$, we get

$$
D_q(c) \geq \frac{1}{c} - \frac{1}{A(q)} + \frac{(1 - 2\log_q c)}{c} \cdot \frac{1}{A(q)}
$$

by letting ε tend to 0. This completes the proof. $\qquad\qquad\square$

Note that for $c < \sqrt{q}$ the above theorem improves the lower bound in Theorem 2.19 by the term

$$
\frac{(1 - 2\log_q c)}{c} \cdot \frac{1}{A(q)}.
$$

We remark that codes providing various forms of traceability such as *traceability codes*, *c-secure frameproof codes* and *identifiable parent property codes*, etc., have been studied by numerous authors (see, for example, [7, 10, 13, 14, 21, 61, 68]).

2.3   Constructions of Perfect Hash Families

Let n and m be integers such that $2 \leq m \leq n$. Let A be a set of size n and let B be a set of size m. A *hash function* is a function h from A to B. We say a hash function $h : A \to B$ is *perfect* on a subset $X \subseteq A$ if h is injective when restricted to X. Let w be an integer such that $2 \leq w \leq m$ and let $\mathcal{H} \subseteq \{h : A \to B\}$.

Definition 2.27. We say $\mathcal{H}$ is an (n, m, w)-*perfect hash family* if for any $X \subseteq A$ with $|X| = w$ there exists at least one function $h \in \mathcal{H}$ such that h is perfect on X. We use $PHF(N; n, m, w)$ to denote an (n, m, w)-perfect hash family with $|\mathcal{H}| = N$

The terminology of "perfect hash family" is motivated by the fact that we have a family of hash functions with the property that if at most w elements are to be hashed, then at least one function in the family yields no collisions when applied to the given w inputs.

Obviously, one can take all functions from A to B which yields a perfect hash family with $|\mathcal{H}| = m^n$. What makes perfect hash families interesting and challenging are the following questions: (i) how can we construct perfect hash families for which $|\mathcal{H}|$ is as small as possible (constructions)?; and (ii) how small can we have $|\mathcal{H}|$ (bounds)?

There have been different definitions and representations of perfect hash families in the literature. For example, a $PHF(N; n, m, w)$ can be depicted as an $N \times n$ array of m symbols, where each row of the array corresponds to one of the functions in the family. This array has the property that, for any subset of w columns, there exists at least one row such that the entries in the w given columns of that row are distinct. A $PHF(N; n, m, w)$ can also be treated as a family of N partitions of an n-set A such that each partition π has at most m parts and such that for all $X \subseteq A$ with $|X| = w$, there exists a partition π for which the elements in X are in distinct parts of π.

Example 2.28 ([2]). The following array gives rise to a $PHF(4; 9, 3, 3)$ which is constructed from a resolvable $(9, 12, 4, 3, 1)$-BIBD (balanced incomplete block design).

1	1	1	2	2	2	3	3	3
1	2	3	1	2	3	1	2	3
1	2	3	3	1	2	2	3	1
1	2	3	2	3	1	3	1	2

Example 2.29 ([6]). Let R be a set of size $r \geq 2$. Set $A = R^3$ and $B = R^2$. We define functions $\phi_1, \phi_2, \phi_3 : A \longrightarrow B$ by

$$\begin{aligned}
\phi_1((a, b, c)) &= (a, b) \\
\phi_2((a, b, c)) &= (b, c) \text{ and} \\
\phi_3((a, b, c)) &= (a, c),
\end{aligned}$$

for all $a, b, c \in R$. Then $\{\phi_1, \phi_2, \phi_3\}$ forms a $PHF(3; r^3, r^2, 3)$.

Perfect hash families originally arose as part of compiler design; see Mehlhorn [39] for a summary of the early results in this area. They have applications to operating systems, language translation systems, hypertext, hypermedia, file managers and information retrieval systems; see the survey article of Czech, Havas and Majewski [15]. More recently, they have found numerous applications to cryptography.

Consider the following example of an application of perfect hash families in threshold cryptography. The main goal of threshold cryptography is to replace a system entity, such as a signer in a classical digital signature scheme, with a group of entities sharing the same power. In a (w, n) threshold signature scheme [16], signature generation requires collaboration of at least w members of a set of

n signers. Although construction of threshold signature schemes generally uses a combination of secret sharing schemes and signature schemes, a simplistic combination of the two primitives could result in a completely insecure system that allows the members of an authorised group to recover the secret key of the signature scheme. In a secure threshold signature scheme, the power of signature generation must be shared among n signers in such a way that w participants can collaborate to produce a valid signature for any given message, whilst no subset of fewer than w signers can forge a signature even if many signatures on different messages are known.

For example, a major problem in the construction of threshold RSA signature schemes is that the secret exponent (key) of RSA, which must be shared among the participants, is an element of $\mathbb{Z}_{\phi(P)}$, which is an abelian group and not a field, where P is the public modulus of RSA and ϕ is Euler's function. This means that the majority of classical secret sharing schemes, such as Shamir's scheme [58], cannot be used directly. A simple and early solution to threshold RSA signature is as follows: a trusted dealer shares the signature key d (the RSA exponent) among n signers $u_1, \ldots, u_n$ such that u_i holds d_i and $d = d_1 + \cdots + d_n$ (mod $\phi(P)$). To sign a message s, each signer u_i produces a partial signature s^{d_i} (mod P). The resulting desired signature is then obtained by multiplication of n partial signatures, as $s^d = s^{d_1} \cdot \ldots \cdot s^{d_n}$ (mod P) for the normal RSA signature on s. This is an (n, n) signature scheme and requires collaboration of every single member of the group for generating a signature. To implement a system with a threshold w, $w < n$, one can generalise the above (n, n) scheme, adopting a protocol such as the one below. The dealer generates the shares of $\binom{n}{w}$ independent runs of a (w, w) additive secret sharing scheme for the same secret, the signing key d in this case, that is for each run i, $1 \leq i \leq \binom{n}{w}$, we have

$$d = d_1^{(i)} + \cdots + d_n^{(i)} \quad (\text{mod } \phi(P)).$$

Then the dealer gives appropriate shares to each signer. Now any w-subset of the group has the complete set of shares for one run of the secret sharing scheme and can sign a message. The main drawback of schemes such as this is their inefficiency, in the sense that each signer has to store shares which are in total $\binom{n-1}{w-1}$ times the size of the RSA signing key. In the following, we show how to apply perfect hash families to improve the above threshold RSA signature scheme.

Let $\mathcal{H} = \{h_1, h_2, \ldots, h_N\}$ be a $PHF(N; n, w, w)$ from $\{1, 2, \ldots, n\}$ to $\{1, 2, \ldots, w\}$. Let d be the RSA secret key which is chosen by the dealer and kept secret. The dealer generates shares of N independent runs of (w, w) secret sharing to share the same secret d, that is,

$$\mathbf{d}^{(i)} = \left(d_1^{(i)}, \ldots, d_w^{(i)} \right),$$

where $d = d_1^{(i)} + \cdots + d_w^{(i)} \pmod{\phi(P)}$ for all $1 \le i \le N$. The dealer then distributes the secret keys to the n signers $u_1, \ldots, u_n$ in such a way that u_ℓ, $1 \le \ell \le n$, holds the keys

$$\hat{\mathbf{d}}_\ell = \left(d_{h_1(\ell)}^{(1)}, d_{h_2(\ell)}^{(2)}, \ldots, d_{h_N(\ell)}^{(N)} \right).$$

Next, generation of the signature is by using the perfect hash family and is reduced to the underlying (w, w) threshold signature scheme. Assume w signers $\{u_i : i \in X\}$, where $X \subseteq \{1, 2, \ldots, n\}$ and $|X| = w$, wish to generate a signature for a message s. From the property of the perfect hash family it follows that there exists a function h_k from $\mathcal{H}$ such that h_k restricted to X is one-to-one. That means, the signers $\{u_i : i \in X\}$ know all the w shares $(d_1^{(k)}, \ldots, d_w^{(k)})$ of the kth runs and so can generate a valid signature using the (w, w) RSA signature as we already explained. On the other hand, each signer holds exactly one share from each run of secret sharing and so for any up to $w-1$ signers, they will have at most $w - 1$ shares for each run and so cannot generate a valid signature, and hence it results in a (w, n) threshold RSA signature.

Now each signer has to store N shares of the RSA signing key, which is N times the size of the underlying RSA secret key. Thus, the complexity of the key storage is dependent on the value of N, the number of functions in $\mathcal{H}$. As we will show, in its optimal form we have perfect hash families with $|\mathcal{H}| = O(\log n)$, which yields a significant improvement on $\binom{n-1}{w-1}$ times the size of the underlying RSA secret key from the trivial solution.

Let $N(n, m, w)$ denote the minimum N for which a $PHF(N; n, m, w)$ exists. We are interested in determining these values. In particular, we are interested in the asymptotic behaviour of $N(n, m, w)$ as a function of n when m and w are fixed.

We first review some bounds for $N(n, m, w)$. All logarithms throughout this subsection are to the base 2. The first result is due to Fredman and Komlós [22].

Theorem 2.30. *We have*

$$N(n, m, w) \ge \frac{\binom{n-1}{w-2} m^{w-2} \log(n - w + 2)}{\binom{m-1}{w-2} n^{w-2} \log(m - w + 2)}.$$

As noted in [5], this lower bound is approximately equal to

$$\frac{m^{w-2}}{m(m-1)(m-2) \cdots (m - (w-1))} \cdot \frac{\log n}{\log(m - w + 2)}$$

as $n \to \infty$ with w and m fixed.

A weaker bound, due to Mehlhorn [39], is as follows.

Theorem 2.31. $N(n, m, w) \geq \frac{\log n}{\log m}$.

This bound can be met when $w = 2$. Indeed, we know that an $N \times n$ array of m symbols is a $PHF(N; n, m, 2)$ if and only if no columns of the array are identical. It follows that there exists an $PHF(N; n, m, 2)$ if and only if $n \leq m^N$. Therefore we obtain explicit constructions such that $N(n, m, 2) = \lceil \frac{\log n}{\log m} \rceil$ for any integers $n \geq m \geq 2$.

Using an elementary probabilistic argument, the following non-constructive upper bound for $N(n, m, w)$ was proved by Mehlhorn [39].

Theorem 2.32. *We have*

$$N(n, m, w) \leq \left\lceil \frac{\log \binom{n}{w}}{\log(m^w) - \log(m^w - w! \binom{m}{w})} \right\rceil.$$

Straightforward approximations using Theorem 2.32 yield the following corollary.

Corollary 2.33 ([39]). $N(n, m, w) \leq \lceil w e^{w^2/m} \log n \rceil$.

From Theorem 2.31 and Corollary 2.33, it follows that for fixed m and w, $N(n, m, w)$ as a function of n is $\Theta(\log n)$. However, the above existence results are non-constructive, and it was believed that it is difficult to give explicit constructions that are asymptotically as good as Corollary 2.33.

Efforts have been made to provide explicit constructions which are much more efficient compared to the trivial solutions or quite reasonable compared to the asymptotically optimal bounds. Most known explicit perfect hash families are constructed from error-correcting codes by Alon [1], resolvable balanced incomplete block designs by Brickell [11] and various inductive techniques by Atici, Magliveras, Stinson and Wei [2]. For a good survey of this subject we refer readers to Blackburn [5].

In [74], Wang and Xing gave a construction of perfect hash families based on function fields over finite fields. The construction proceeds as follows.

Let T be a nonempty set of rational places of $F/\mathbb{F}_q$. Let G be a divisor of F with $T \cap \text{Supp}(G) = \emptyset$. As usual, let $\mathcal{L}(G)$ be the vector space formed by

$$\mathcal{L}(G) = \{f \in F \setminus \{0\} : \text{div}(f) + G \geq 0\} \cup \{0\}.$$

Let each place $P \in T$ be associated with a map h_P from $\mathcal{L}(G)$ to $\mathbb{F}_q$ defined by $h_P(f) = f(P)$. Put $\mathcal{H} = \{h_P : P \in T\}$. It is shown as in the proof of Lemma 2.7 that if $\deg(G) \geq 2g(F) + 1$, then the cardinality of $\mathcal{H}$ is equal to $|T|$.

Theorem 2.34 ([74]). *Let $F/\mathbb{F}_q$ be a global function field and T a nonempty set of rational places of F. Suppose that G is a divisor of F with $\deg(G) \geq$*

$2g(F)+1$ and $T \cap \text{Supp}(G) = \emptyset$. Then $\mathcal{H}$ is a $PHF(|T|; q^{\deg(G)-g(F)+1}, q, w)$ provided that $2 \le w \le q$ and $|T| > \deg(G) \cdot \binom{w}{2}$.

Proof. Let $\mathcal{H}$ be as the above. For a subset X of $\mathcal{L}(G)$ with w elements, consider the set

$$\mathbf{S}_X := \{(u - v)^2 : u, v \in X, u \ne v\}.$$

Then $\mathbf{S}_X$ has at most $\binom{w}{2}$ elements and the number of zeros of an element $(u - v)^2$ is equal to the number of zeros of $u - v$. The number of zeros of $u - v$ from the set T is at most $\deg(G)$ since $u - v$ is an element of $\mathcal{L}(G)$ and $T \cap \text{Supp}(G) = \emptyset$. Therefore, the number of zeros from T of all functions in $\mathbf{S}_X$ is at most

$$\deg(G) \cdot |\mathbf{S}_X| \le \deg(G) \cdot \binom{w}{2}.$$

By the condition $|T| > \deg(G) \cdot \binom{w}{2}$, we can find a place $R \in T$ such that R is not a zero for any element of $\mathbf{S}_X$.

We claim that the function h_R is one-to-one on the subset X. Indeed, suppose u and v are two different elements of X. Then $(u - v)^2 \in \mathbf{S}_X$, thus R is not a zero of $(u - v)^2$, i.e., $u(R) \ne v(R)$. This is equivalent to $h_R(u) \ne h_R(v)$. The proof is complete. □

Theorem 2.34 gives a construction of perfect hash families based on general global function fields. Applying Theorem 2.34 to the function fields due to Garcia and Stichenoth (see Chapter 1), we obtain perfect hash families with nice parameters:

Let q be a square and put $r = \sqrt{q}$. Consider a sequence of function fields $F_i/\mathbb{F}_q$ as follows (see Chapter 1, Section 5.1). Let F_1 be the rational function field $F_1 = \mathbb{F}_q(x_1)$. For $i \ge 2$ let $F_i = F_{i-1}(x_i)$ with

$$x_i^r + x_i = \frac{x_{i-1}^r}{x_{i-1}^{r-1} + 1}.$$

Then the number of rational places of F_i is more than $(r - 1)r^i$ and the genus g_i of F_i is less than r^i for all $i \ge 1$.

Put

$$N_i = (\sqrt{q} - 1)q^{i/2}, \quad t_i = \lfloor (c+1)q^{i/2} \rfloor, \quad w = \lfloor \sqrt{\frac{2}{c+1}} q^{1/4} \rfloor,$$

where $1 \le c \le (\sqrt{q} - 2)/2$ is a real constant independent of i. Then

$$N_i > t_i \binom{w}{2} \qquad \text{for all } i \ge 1,$$

and there exist a subset T_i of rational places of F_i with $|T_i| = N_i$ and a divisor G_i of F_i of degree t_i such that $T_i \cap \mathrm{Supp}(G_i) = \emptyset$. Applying Theorem 2.34 gives a $PHF(N_i; q^{t_i+1-g_i}, q, w)$ for all $i \geq 1$. Since $t_i+1-g_i > \lfloor (c+1)r^i \rfloor +1-r^i > \lfloor cq^{i/2} \rfloor$, we obtain the following result.

Theorem 2.35. *Let the prime power $q \geq 16$ be a square and let c be a real number with $1 \leq c \leq (\sqrt{q} - 2)/2$. Then there exists a*

$$PHF\left((\sqrt{q}-1)q^{i/2}; q^{\lfloor cq^{i/2}\rfloor}, q, \left\lfloor \sqrt{\frac{2}{c+1}}q^{1/4} \right\rfloor \right)$$

for each $i \geq 1$. In particular, taking $c = 1$, we obtain a

$$PHF\left((\sqrt{q}-1)q^{i/2}; q^{q^{i/2}}, q, \lfloor q^{1/4}\rfloor \right)$$

for each $i \geq 1$.

Before describing the next result, we first recall a simple composition construction of perfect hash families, due to Blackburn, Burmester, Desmedt and Wild [8, 5]. Assume that $\mathcal{H}_1$ is a $PHF(N; n, n_0, w)$ from A_1 to B_1 and $\mathcal{H}_2$ is a $PHF(N_0; n_0, m, w)$ from A_2 to B_2 such that $B_1 = A_2$. Then it is fairly straightforward to verify that $\mathcal{H} = \{h_2 h_1 : h_1 \in \mathcal{H}_1, h_2 \in \mathcal{H}_2\}$ is a $PHF(NN_0; n, m, w)$ from A_1 to B_2. That is, we have the following result.

Lemma 2.36 ([5]). *Suppose there exist explicit hash families $PHF(N; n, n_0, w)$ and $PHF(N_0; n_0, m, w)$. Then there exists an explicit $PHF(NN_0; n, m, w)$.*

Combining the above composition method with Theorem 2.35, we arrive at the following result from [74].

Theorem 2.37 ([74]). *For any integers $m \geq w \geq 2$, there exist explicit constructions of $PHF(N; n, m, w)$ such that $N \leq C \log n$, where C is a constant independent of n and n can go to ∞.*

Proof. Let q be the least square prime power such that $q \geq m$ and $q^{1/4} \geq w$. Since $g(x) = \sqrt{\frac{2}{x+1}}q^{1/4}$ is continuous on $[1, \infty)$, it follows that we may choose a $c_0 \in [1, \infty)$ such that $\sqrt{\frac{2}{c_0+1}}q^{1/4} = w$. Then $1 \leq c_0 \leq (\sqrt{q} - 2)/2$ as required in Theorem 2.35. By Theorem 2.35, we know that there is an explicit $PHF((\sqrt{q}-1)q^{i/2}; q^{\lfloor c_0 q^{i/2}\rfloor}, q, w)$ for all $i \geq 1$. Since $q \geq m$, there exists an explicit $PHF(N_0; q, m, w)$, where the parameter N_0 can be effectively determined by m because of the previous choice of q. From Lemma 2.36, it follows that there exist constructions for

$$PHF(N_0(\sqrt{q}-1)q^{i/2}; q^{\lfloor c_0 q^{i/2}\rfloor}, m, w)$$

for all $i \geq 1$. We thus obtain $PHF(N; n, m, w)$ with $N \leq C \log n$, where

$$C \approx \frac{N_0(\sqrt{q} - 1)}{c_0 \log q}$$

in which all the parameters on the right-hand side depend only on m and w, but n can go to ∞ as $n = q^{\lfloor c_0 q^{i/2} \rfloor}$ for all $i = 1, 2, \ldots$. The desired result follows.   □

Finally we remark that in recent years perfect hash families have found numerous applications to cryptography, for example, to broadcast encryption [20], secret sharing [8], key distribution patterns [67], anti-jamming radio networks [17], distributing the encryption and decryption of block ciphers [38], secure multicasting [53] and frameproof codes [61, 68].

On the other hand, a great deal of research efforts on constructing perfect hash families with good parameters and deriving bounds on $N(n, m, w)$ have been carried out. In [2], Atici, Magliveras, Stinson and Wei provided various recursive methods resulting in explicit constructions of $PHF(N; n, m, w)$ in which N is a polynomial function of $\log n$ for fixed m and w. Stinson, Wei and Zhu [69] employed some combinatorial techniques to generalise and improve results from [2]. For given m and w, they constructed $PHF(N; n, m, w)$ in which N is $O(C^{\log^*(n)} \log n)$, where C is a constant depending only on w and $\log^*$ is the function from $\mathbb{Z}^+$ to $\mathbb{Z}^+$ recursively defined as follows:

$$\log^*(1) = 1, \qquad \log^*(n) = \log^*(\lceil \log n \rceil) + 1 \quad \text{for } n > 1.$$

Blackburn and Wild [9] introduced and studied *linear* perfect hash families and showed that there exist explicit constructions for $PHF(N; n, m, w)$ in which $N = (w - 1)(\log n)/\log m$, where m is a prime power and n is a power of m.

2.4   Constructions of Cover-Free Families

Definition 2.38. A set system $(X, \mathcal{B})$ with $X = \{x_1, \ldots, x_t\}$ and $\mathcal{B} = \{B_i \subseteq X : i = 1, \ldots, n\}$ is called an (n, t, r)-*cover-free family* (or (n, t, r)-CFF for short) if for any subset $\Delta \subseteq \{1, \ldots, n\}$ with $|\Delta| = r$ and any $i \in \{1, \ldots, n\} \backslash \Delta$ we have

$$\left| B_i \backslash \bigcup_{j \in \Delta} B_j \right| \geq 1.$$

We say $(X, \mathcal{B})$ is k-*uniform* if $|B_i| = k$ for all $1 \leq i \leq n$.

Note that an $(n, t, 1)$-CFF is exactly a Sperner system. Another trivial cover-free family is the family consisting of single-element subsets, in which $n = t$. We will be interested in (n, t, r)-CFF for which $n > t$, in particular, for given t and r, we want the value of n to be as large as possible.

Cover-free families were first studied in terms of superimposed binary codes by Kautz and Singleton [33] in 1964. These codes are related to files retrieval, data communication and magnetic memories. In 1985 Erdös, Frankl and Füredi [19] studied cover-free families as combinatorial objects, generalising the Sperner systems. Since then, they have been discussed by numerous researchers in the subjects of information theory, combinatorics, communication and cryptography.

Let us start with an example of an application of cover-free families in multi-receiver authentication systems. Conventional authentication systems, as discussed in Section 2.1, deal with *point-to-point* message authentication in which the sender and the receiver share a secret key and are both assumed honest. Multi-receiver authentication systems are an extension of the point-to-point authentication model in which there are multiple receivers who cannot all be trusted. The sender broadcasts a message to all the receivers who can individually verify authenticity of the message using their secret key information. There are malicious groups of receivers who use their secret keys and all the previous communications in the system to construct fraudulent messages. They succeed in their attack as soon as a single receiver accepts the message as being authentic. In an (r, n) multi-receiver authentication system there are n receivers such that the coalition of any r receivers cannot cheat other receivers.

Obviously, a multi-receiver authentication system can be constructed from a conventional authentication code by allowing the sender to use n authentication keys for the n receivers and broadcast a codeword that is simply a concatenation of the codewords for each receiver. The length of the combined authentication tag is n times the length of the individual receivers' authentication tags, and the sender's key is n times the size of a receiver's key. This is a very uneconomical method of authenticating a message as such a system can prevent attacks by even $n - 1$ colluding receivers, while it is reasonably realistic to assume that an (r, n) multi-receiver authentication system is sufficient to satisfy the security requirements and r is a system parameter independent of n. That is, we assume that in every group of $r + 1$ receivers there is at least one honest receiver. In the following we show that cover-free families can play a role to improve the above trivial construction.

Assume that $(X, \mathcal{B})$ is an (n, t, r)-CFF and $(\mathcal{S}, \mathcal{E}, \mathcal{T})$, together with the authentication mapping $f : \mathcal{S} \times \mathcal{E} \to \mathcal{T}$, is an authentication code [1]. We construct an (r, n) multi-receiver authentication system with n receivers $R_1, \ldots, R_n$ as follows. The sender randomly chooses a t-tuple of keys $(e_1, \ldots, e_t) \in \mathcal{E}^t$ and privately sends e_i to every receiver R_j for all j with $x_i \in B_j$, $1 \leq i \leq t$. To authenticate a source message $s \in \mathcal{S}$, the sender computes $a_i = f(s, e_i)$

---

[1]$(\mathcal{S}, \mathcal{E}, \mathcal{T})$ can be either an unconditionally secure A-code or a computationally secure MAC.

for all $1 \le i \le t$ and broadcasts $(s, a_1, \ldots, a_t)$ to all the receivers. Since the receiver R_j holds the keys e_i for all i with $x_i \in B_j$, R_j accepts $(s, a_1, \ldots, a_m)$ as authentic if $a_i = f(s, e_i)$ for all i satisfying $x_i \in B_j$.

It has been proved in [51] that this construction gives rise to an (r, n) multi-receiver authentication system where both the sizes of key for the sender and of the broadcasting message are t times increased from the underlying point-to-point authentication codes, in contrast to the n times increase in the trivial construction.

It is clear from the above example that we want to maximise the value of n when t and r are fixed, or equivalently, to minimise the value of t while n and r are given. It is also easy to see that there is a trade-off between the values of n and t in a cover-free family.

As we mentioned before, a Sperner system is an $(n, t, 1)$-CFF in which from Sperner's Theorem we know that $n \le \binom{t}{\lfloor \frac{t}{2} \rfloor}$ and the bound is tight.

In [19], Erdös, Frankl and Füredi gave an upper bound on n for uniform cover-free families.

Theorem 2.39 ([19]). *In a k-uniform (n, t, r)-CFF we have*

$$n \le \binom{t}{\lceil \frac{k}{r} \rceil} \bigg/ \binom{k}{\lceil \frac{k}{r} - 1 \rceil} .$$

As pointed out by Wei [77], the above bound can be reached for some special cases. For example, there exists a probabilistic construction in [19] for a $2r$-uniform (n, t, r)-CFF with $n = (t^2/(4r)) - o(t^2)$. However, it remains open whether or not the bound in Theorem 2.39 is the best possible bound for the uniform case.

In the general case, the best lower bound on t is given in [70]. Proofs of variants of the following theorem can be found in [18, 23, 50].

Theorem 2.40. *For any (n, t, r)-CFF with $r \ge 2$, we have*

$$t \ge c \frac{r^2}{\log r} \log n$$

for some constant $c \approx 1/2$.

On the other hand, using a probabilistic method, Erdös, Frankl and Füredi [19] proved the existence of (n, t, r)-CFF with $t = O(r^2 \log n)$ and $|B_i| = O(r \log n)$ for any $r \ge 2$. Explicit constructions to achieve this bound asymptotically are of high interest.

We first give two constructions of cover-free families from perfect hash families.

Construction from PHF I: This gives a direct construction of cover-free families from perfect hash families. Assume that $\mathcal{H}$ is a $PHF(N; n, m, w)$ from A to B. Let $A = \{1, 2, \ldots, n\}$ and $B = \{1, 2, \ldots, m\}$. We define

$$X = \mathcal{H} \times B = \{(h, j) : h \in \mathcal{H}, j \in B\}.$$

For each $1 \leq i \leq n$, we define a subset (block) B_i of X by

$$B_i = \{(h, h(i)) : h \in \mathcal{H}\},$$

and $\mathcal{B} = \{B_i : 1 \leq i \leq n\}$. Then $(X, \mathcal{B})$ is an $(n, Nm, w - 1)$-CFF. Clearly, $|X| = Nm$ and $|\mathcal{B}| = n$. For any w blocks $B_{i_1}, \ldots, B_{i_w}$, since $\mathcal{H}$ is a $PHF(N; n, m, w)$, there exists a perfect hash function $h \in \mathcal{H}$ such that h restricted to $\{i_1, \ldots, i_w\}$ is one-to-one. It follows that $h(i_1), \ldots, h(i_w)$ are w distinct elements in B, which also implies that $(h, h(i_1)), \ldots, (h, h(i_w))$ are w distinct elements in $B_{i_1}, \ldots, B_{i_w}$, respectively. So the union of any $w - 1$ blocks in $\mathcal{B}$ cannot cover any remaining block. Thus, we have shown the following result.

Theorem 2.41. *If there exists a $PHF(N; n, m, w)$, then there exists an $(n, Nm, w - 1)$-CFF.*

Applying Theorem 2.37 and its proof, we immediately obtain the following result.

Corollary 2.42. *For any integer $r \geq 2$, there exists an explicit construction for (n, t, r)-CFF in which t is $O(r^4 \log n)$.*

Construction from PHF II: This construction provides a method of building new cover-free families from old ones, using perfect hash families. The construction works as follows. Let $(X_0, \mathcal{B}_0)$ be an $(n_0, t_0, w - 1)$-CFF and let $\mathcal{H} = \{h_1, \ldots, h_N\}$ be a $PHF(N; n, n_0, w)$. Consider N copies of $(X_0, \mathcal{B}_0)$, denoted by $(X_1, \mathcal{B}_1), \ldots, (X_N, \mathcal{B}_N)$, where X_i and X_j are disjoint sets, i.e. $X_i \cap X_j = \emptyset$, for all $i \neq j$. For each $1 \leq j \leq N$, let $X_j = \{x_1^{(j)}, \ldots, x_{t_0}^{(j)}\}$ and $\mathcal{B}_j = \{B_1^{(j)}, \ldots, B_{n_0}^{(j)}\}$. Then $(X_j, \mathcal{B}_j)$ is an $(n_0, t_0, w - 1)$-CFF. We construct a pair $(X, \mathcal{B})$ with

$$X = X_1 \cup \cdots \cup X_N \quad \text{and} \quad \mathcal{B} = \{B_1, \ldots, B_n\},$$

where $B_i = B_{h_1(i)}^{(1)} \cup \cdots \cup B_{h_N(i)}^{(N)} = \cup_{j=1}^{N} B_{h_j(i)}^{(j)}$ for $1 \leq i \leq n$. That is, an element of $\mathcal{B}$ is a union of elements of $\mathcal{B}_j$, $1 \leq j \leq N$, chosen through the application of a perfect hash family. We show that $(X, \mathcal{B})$ is an $(n, Nt_0, w - 1)$-CFF. Clearly, $|X| = Nt_0$ and $|\mathcal{B}| = n$. Now for any given w blocks $\{B_i, B_{j_1}, \ldots, B_{j_{w-1}}\} \subseteq \mathcal{B}$, there exists at least one hash function $h_k \in \mathcal{H}$ which is one-to-one on $\{i, j_1, \ldots, j_{w-1}\}$. For each $1 \leq j \leq N$, consider the set system $(X_j, \mathcal{B}_j)$, which may be regarded as a subsystem of $(X, \mathcal{B})$. Since $(X_k, \mathcal{B}_k)$ is an $(n_0, t_0, w-1)$-CFF, we have

$$|B_i \setminus (B_{j_1} \cup \cdots \cup B_{j_{w-1}})| \geq |B^{(k)}_{h_k(i)} \setminus (B^{(k)}_{h_k(j_1)} \cup \cdots \cup B^{(k)}_{h_k(j_{w-1})})|$$
$$\geq 1,$$

which proves the desired result. Thus, we have shown the following theorem.

Theorem 2.43. *Suppose that there exists an $(n_0, t_0, w-1)$-CFF and a $PHF(N; n, n_0, w)$. Then there exists an $(n, Nt_0, w-1)$-CFF.*

Construction from error-correcting codes: Another nice construction of cover-free families is through error-correcting codes. Let Y be an alphabet of q elements. Recall that an (N, M, d, q) *code* is a set $\mathcal{C}$ of M vectors in Y^N such that the Hamming distance between any two distinct vectors in $\mathcal{C}$ is at least d.

Consider an (N, M, d, q) code $\mathcal{C}$. We write each codeword as $c_i = (c_{i1}, \ldots, c_{iN})$ with $c_{ij} \in Y$, where $1 \leq i \leq M, 1 \leq j \leq N$. Set $X = \{1, \ldots, N\} \times Y$ and $\mathcal{B} = \{B_i : 1 \leq i \leq M\}$, where for each $1 \leq i \leq M$ we define $B_i = \{(j, c_{ij}) : 1 \leq j \leq N\}$. It is easy to see that $|X| = Nq$, $|\mathcal{B}| = M$ and $|B_i| = N$. For each choice of $i \neq k$, we have $|B_i \cap B_k| = |\{(j, c_{ij}) : 1 \leq j \leq N\} \cap \{(j, c_{kj}) : 1 \leq j \leq N\}| = |\{j : c_{ij} = c_{kj}\}| \leq N - d$.

It is straightforward to show that $(X, \mathcal{B})$ is an (M, Nq, r)-CFF if the condition $r < \frac{N}{N-d}$ holds. We thus obtain the following theorem.

Theorem 2.44. *If there is an (N, M, d, q) code, then there exists an (M, Nq, r)-CFF provided that $r < \frac{N}{N-d}$.*

Now if we apply the above code construction to algebraic-geometry codes, we immediately obtain the following corollary.

Corollary 2.45 ([48]). *Let $F/\mathbb{F}_q$ be a global function field of genus g with $N+1$ rational places. Then for any integer $\ell \geq 1$ with $g \leq \ell < N$, there exists a $(q^{\ell-g+1}, Nq, \lfloor (N-1)/\ell \rfloor)$-CFF.*

We are interested in the asymptotic behaviour of cover-free families in Corollary 2.45. We use the quantity $A(q)$ introduced in (2.4). By Ihara [30] and Vlăduţ and Drinfeld [73], we know that $A(q) = \sqrt{q} - 1$ if q is a square prime power (see also Chapter 1). Thus, we have the following asymptotic result.

Corollary 2.46. *For a fixed $r \geq 1$ and a square prime power q with $r < \sqrt{q}-1$, there exists a sequence of CFFs with parameters*

$$(q^{\ell_i-g+1}, N_i q, r)$$

such that

$$\lim_{i \to \infty} \frac{\log q^{\ell_i-g+1}}{N_i q} = \frac{\log q}{q} \cdot \left(\frac{1}{r} - \frac{1}{\sqrt{q}-1}\right). \tag{2.6}$$

Corollary 2.46 shows that for any fixed r there are infinite families of (n, t, r)-CFFs in which $t = O(\log n)$ and the constructions are explicit.

Next, we improve the asymptotic result in Corollary 2.46 by applying the function field construction directly.

Let $F/\mathbb{F}_q$ be a global function field of genus g with at least $m+g+1$ rational places. By using an argument similar to that in the proof of [62, Proposition I.6.10], we can show that there exist $m + 1$ rational places $P_\infty, P_1, \ldots, P_m$ of F such that

$$\mathcal{L}(r\ell P_\infty - \sum_{i=1}^{m} P_i) = \{0\},$$

provided that $r\ell - m \leq g - 1$.

For each $f \in \mathcal{L}(\ell P_\infty)$, we denote $B_f = \{(P_i, f(P_i)) : i = 1, 2, \ldots, m\}$. For any $r + 1$ distinct $f_1, f_2, \ldots, f_r, f \in \mathcal{L}(\ell P_\infty)$, we have $\prod_{i=1}^{r}(f - f_i) \in \mathcal{L}(\ell P_\infty)$ as $f \neq f_i$ for any $i = 1, \ldots, r$. On the other hand, since $\prod_{i=1}^{r}(f - f_i) \neq 0$, we have $\prod_{i=1}^{r}(f - f_i) \notin \mathcal{L}(r\ell P_\infty - \sum_{i=1}^{m} P_i)$. Therefore, there exists a P_j such that $\prod_{i=1}^{r}(f - f_i)(P_j) \neq 0$, i.e., $f(P_j) \neq f_i(P_j)$, $\forall i = 1, 2, \ldots, r$. We then conclude $(P_j, f(P_j)) \in B_f$ and $(P_j, f(P_j)) \notin \cup_{i=1}^{r} B_{f_i}$. That is,

$$|B_f \setminus \cup_{i=1}^{r} B_{f_i}| \geq 1.$$

This shows that there exists a (q^{l-g+1}, mq, r)-CFF. Moreover, if we let $r = \lfloor g - 1 + m/\ell \rfloor$, then we obtain a

$$(q^{\ell-g+1}, mq, \lfloor g - 1 + m/l \rfloor)\text{-CFF}.$$

Thus, we obtain the following result.

Corollary 2.47 ([48]). *If q is a square prime power, then for a fixed $r \geq 1$ we obtain a sequence of CFFs with parameters*

$$(q^{\ell_i - g + 1}, N_i q, r)$$

such that

$$\lim_{i \to \infty} \frac{\log q^{\ell_i - g + 1}}{N_i q} = \frac{\log q}{q} \cdot \left(\frac{1}{r} - (1 - \frac{1}{r}) \frac{1}{\sqrt{q} - 2} \right). \tag{2.7}$$

Obviously, the bound (2.7) improves the bound (2.6) for $r < \sqrt{q} - 1$. We note that Corollaries 2.46 and 2.47 show that the existence bounds for CFFs in [19, 70] can be asymptotically met by the explicit constructions.

We conclude this section by mentioning a generalisation of cover-free families, called (s, r)-cover-free families, defined below.

Definition 2.48. Let X be a set of t elements (points) and let $\mathcal{B}$ be a set of n subsets (blocks) of X. Then $(X, \mathcal{B})$ is called an (s, r)-*cover-free family* provided that, for any s blocks $B_1, \ldots, B_s$ in $\mathcal{B}$ and r other blocks $B'_1, \ldots, B'_r$ in $\mathcal{B}$, one has

$$\cap_{i=1}^{s} B_i \not\subseteq \cup_{j=1}^{r} B'_j.$$

In words, no intersection of up to s blocks is contained in the union of r other blocks. For $s = 1$, a $(1, r)$-cover-free family is exactly the one we studied earlier in this subsection. For $s = 2$, a $(2, w)$-cover-free family was introduced, under the name of *key distribution pattern*, by Mitchell and Piper [40] to provide a mechanism for distributing a secret key to each pair of users in a network. For general s, they are relevant to conference key distribution and broadcast encryption [20, 66].

Cover-free families and their generalisations have been used for many other cryptographic problems such as frameproof codes and traceability schemes [21, 61, 68], authentication for group communication [51, 53], broadcast anti-jamming systems [17], multiple time signature schemes [48] and blacklisting problems [35], to just mention a few.

3. Applications to Stream Ciphers and Linear Complexity

3.1 Background

A *cryptosystem* protects sensitive data by transforming the original data (i.e., the *plaintext*) into encrypted data (i.e., the *ciphertext*) and allowing unique recovery of the plaintext from the ciphertext by decryption. The encryption and decryption algorithms depend on the choice of parameters called *keys*, with the provision that the number of possible keys is so large that it inhibits exhaustive key search by an attacker.

A *stream cipher* is a symmetric (or private-key) cryptosystem since it uses the same key for encryption and decryption. This common key has to be kept secret. In practical implementations, a stream cipher will be a bit-based cryptosystem. The plaintext, the ciphertext and the key are all bit strings of the same length, but this length can be arbitrary (as opposed to a block cipher where the lengths are fixed). Encryption proceeds by taking the plaintext string and bitwise XORing it with the key string (or, in other words, adding the two strings bit by bit in the finite field $\mathbb{F}_2$). It is clear that the plaintext is recovered by bitwise XORing the ciphertext string and the key string. Thus, in a bit-based stream cipher the encryption and decryption algorithms are identical, which has the practical advantage that the same hardware can be used for both operations.

From the theoretical point of view, it does not make any difference whether we consider stream ciphers over $\mathbb{F}_2$ or over an arbitrary finite field $\mathbb{F}_q$. Therefore we consider the general case where plaintext and ciphertext are strings (or, in other words, finite sequences) of elements of $\mathbb{F}_q$ and encryption and decryption proceed by termwise addition, respectively subtraction, of the same key string of elements of $\mathbb{F}_q$. The key string is commonly called the *keystream* and known only to authorised users. It is convenient from now on to speak of strings, respectively sequences, over $\mathbb{F}_q$ when we mean strings, respectively sequences, of elements of $\mathbb{F}_q$.

In an ideal situation, the keystream would be a "truly random" string over $\mathbb{F}_q$. In this case, the stream cipher would be perfectly secure since the ciphertexts will carry absolutely no information, and so there will be no basis for an attack on the cryptosystem. In practice, sources of true randomness are hard to come by, and so keystreams are taken to be pseudorandom strings that are obtained from certain secret seed data by a (perhaps even publicly available) algorithm. A central issue in the security analysis of stream ciphers is then the quality assessment of these pseudorandom keystreams. In other words, we need to know how close a given keystream is to true randomness. We focus here on the complexity-theoretic aspects of this assessment where global function fields have played some methodological role. There are also statistical techniques for this assessment which will not concern us here. For general background on stream ciphers we refer to Chapter 4 and to the survey article of Rueppel [49].

3.2 Linear Complexity

Most practical keystreams are built up from linear recurring sequences, that is, sequences generated by linear recurrence relations. Hence it is a natural idea to look for the linear recurrence relation of lowest order that can generate the keystream. Clearly, if a keystream can be generated by a low-order linear recurrence relation, then it is easily predictable, hence distinctly nonrandom, and thus has to be discarded as unsuitable. This viewpoint leads to the following notions of complexity.

Definition 3.1. Let n be a positive integer and let S be a sequence over $\mathbb{F}_q$. Then the *nth linear complexity* $L_n(S)$ of S is the least k such that the first n terms of S can be generated by a linear recurrence relation over $\mathbb{F}_q$ of order k. The *linear complexity* $L(S)$ of the sequence S is defined by

$$L(S) = \sup_{n \geq 1} L_n(S).$$

It is clear that we always have $0 \leq L_n(S) \leq n$ and $L_n(S) \leq L_{n+1}(S)$. The extreme values of $L_n(S)$ correspond to highly nonrandom behaviour, for if $s_1, \ldots, s_n$ are the first n terms of S, then $L_n(S) = 0$ if and only if $s_i = 0$ for $1 \leq i \leq n$, whereas $L_n(S) = n$ if and only if $s_i = 0$ for $1 \leq i \leq n-1$ and $s_n \neq 0$. Note also that if S is ultimately periodic, then we have $L(S) < \infty$. For arbitrary sequences over $\mathbb{F}_q$ we have the following concept.

Definition 3.2. For a sequence S over $\mathbb{F}_q$, let $L_n(S)$ denote again the nth linear complexity of S. Then the sequence $L_1(S), L_2(S), \ldots$ is called the *linear complexity profile* of S.

The linear complexity profile of S is a nondecreasing sequence of nonnegative integers. Thus, the linear complexity profile is fully determined if we know where and how large its jumps are. These data can be conveniently described in terms of the continued fraction expansion of the generating function of S. If $s_1, s_2, \ldots$ are the terms of S, then the *generating function* of S is given by

$$G := \sum_{i=1}^{\infty} s_i x^{-i} \in \mathbb{F}_q((x^{-1})).$$

Here $\mathbb{F}_q((x^{-1}))$ is the field of formal Laurent series over $\mathbb{F}_q$ in the variable x^{-1}, or equivalently $\mathbb{F}_q((x^{-1}))$ is the completion of the rational function field $\mathbb{F}_q(x)$ with respect to its infinite place. The continued fraction expansion of G has the form

$$G = 1/(A_1 + 1/(A_2 + \cdots)),$$

where the partial quotients $A_1, A_2, \ldots$ are polynomials over $\mathbb{F}_q$ of positive degree. This expansion is finite if G is rational, i.e., if $G \in \mathbb{F}_q(x)$, and infinite if G is irrational, i.e., if $G \notin \mathbb{F}_q(x)$. It can be shown that the jumps in the linear complexity profile of S are exactly the degrees $\deg(A_1), \deg(A_2), \ldots$ of the partial quotients and that the locations of the jumps are also uniquely determined by these degrees (see [46, Section 7.1]).

Let μ_q be the uniform probability measure on $\mathbb{F}_q$ which assigns the measure $1/q$ to each element of $\mathbb{F}_q$. Let $\mathbb{F}_q^\infty$ denote the sequence space over $\mathbb{F}_q$ and let μ_q^∞ be the complete product probability measure on $\mathbb{F}_q^\infty$ induced by μ_q. Then, except on a set of sequences S with μ_q^∞-measure 0, we have

$$\lim_{n \to \infty} \frac{L_n(S)}{n} = \frac{1}{2}. \tag{3.1}$$

We refer to [46, Section 7.1] for an elementary proof of this fact. This result suggests to study the deviations of $L_n(S)$ from $n/2$. There has been a strong interest in sequences S for which these deviations are bounded.

Definition 3.3. Let d be a positive integer. Then a sequence S over $\mathbb{F}_q$ is called *d-perfect* if

$$|2L_n(S) - n| \le d \qquad \text{for all } n \ge 1.$$

A 1-perfect sequence is also called *perfect*. A sequence is called *almost perfect* if it is d-perfect for some d.

An alternative characterisation of d-perfect sequences can be based on continued fractions (see [46, Theorem 7.2.2]).

Proposition 3.4. *A sequence S over $\mathbb{F}_q$ is d-perfect if and only if the generating function G of S is irrational and the partial quotients A_j in the continued fraction expansion of G satisfy $\deg(A_j) \leq d$ for all $j \geq 1$.*

Example 3.5. Let $q = 2$ and let S be the sequence $s_1, s_2, \ldots$ over $\mathbb{F}_2$ defined by $s_i = 1$ if $i = 2^h - 1$ for some integer $h \geq 1$ and $s_i = 0$ otherwise. Then the generating function G of S satisfies the identity $G^2 = xG + 1$ in $\mathbb{F}_2((x^{-1}))$, hence $G = x + G^{-1}$. This leads to a continued fraction expansion of G with partial quotients $A_j = x$ for all $j \geq 1$. Therefore the sequence S is perfect by Proposition 3.4.

The following convenient sufficient condition for a sequence to be d-perfect can be obtained from [46, Remark 7.2.3].

Proposition 3.6. *Let S be a sequence over $\mathbb{F}_q$ and let d be a positive integer. If*

$$L_n(S) \geq \frac{n+1-d}{2} \qquad \text{for all } n \geq 1,$$

then S is d-perfect.

3.3   Constructions of Almost Perfect Sequences

The idea of using global function fields for the construction of almost perfect sequences is due to Xing and Lam [84]. Throughout this subsection, we let $F/\mathbb{F}_q$ be a global function field, P a rational place of F, and t a local parameter at P with $\deg((t)_\infty) = 2$.

We first describe the construction of Xing and Lam [84]. Let $f \in F$ with $f \notin \mathbb{F}_q(t)$ and $\nu_P(f) \geq 0$. Then the local expansion of f at P has the form

$$f = \sum_{i=0}^{\infty} s_i t^i$$

with all $s_i \in \mathbb{F}_q$. From this expansion we read off the sequence S_1 of coefficients $s_1, s_2, \ldots$.

Theorem 3.7. *If the integer d is such that $d \geq \deg((f)_\infty)$, then the sequence S_1 constructed above is d-perfect.*

Proof. We proceed by Proposition 3.6. Fix $n \geq 1$, put $k = L_n(S_1)$ and write a linear recurrence relation over $\mathbb{F}_q$ of order k satisfied by $s_1, \ldots, s_n$ in the form

$$\sum_{h=0}^{k} a_h s_{i+h} = 0 \qquad \text{for } 1 \leq i \leq n - k, \tag{3.2}$$

where $a_h \in \mathbb{F}_q$ for $0 \le h \le k$ and $a_k = 1$. Consider

$$b := f \sum_{h=0}^{k} a_{k-h} t^h - \sum_{j=0}^{k} \left(\sum_{i=0}^{j} a_{k-j+i} s_i \right) t^j.$$

Note that b is a nonzero element of F since $a_k \ne 0$ and $f \notin \mathbb{F}_q(t)$. By applying the linear recurrence relation (3.2) and considering the local expansion of b at P, we obtain $\nu_P(b) \ge n + 1$. On the other hand, the pole divisor of b satisfies

$$(b)_\infty \le (f)_\infty + (t^k)_\infty.$$

Therefore

$$n + 1 \le \nu_P(b) \le \deg((b)_0) = \deg((b)_\infty) \le d + 2k,$$

and so

$$k \ge \frac{n + 1 - d}{2}.$$

Thus, S_1 is d-perfect by Proposition 3.6. $\qquad\qquad\square$

Example 3.8. Let $q = 2$, let F be the rational function field $\mathbb{F}_2(x)$, and let P be the zero of x. We choose $t = x^2 + x$ and $f = x/(x + 1)$. Then at P we have the local expansion

$$f = \frac{x^2}{t} = \sum_{h=1}^{\infty} t^{2^h - 1}.$$

The sequence S_1 of coefficients $1, 0, 1, 0, 0, 0, 1, \ldots$ is perfect by Theorem 3.7. Note that this is the same sequence as in Example 3.5.

The case where $f \in F$ with $f \notin \mathbb{F}_q(t)$ and $\nu_P(f) < 0$ can be reduced to that in Theorem 3.7, as was pointed out by Xing *et al.* [87]. Indeed, if we put $w = -\nu_P(f) > 0$, then $\nu_P(t^w f) = 0$ and we have a local expansion at P of the form

$$t^w f = \sum_{i=0}^{\infty} s_i t^i$$

with all $s_i \in \mathbb{F}_q$. From this expansion we read off the sequence S_1' of coefficients $s_1, s_2, \ldots$.

Corollary 3.9. *If $\nu_P(f) = -w < 0$ and the integer d is such that $d \ge \deg((f)_\infty)$, then the sequence S_1' constructed above is $(d + w)$-perfect.*

Proof. We have

$$\deg((t^w f)_\infty) \le \deg((f)_\infty) + \deg((t^w)_\infty) - \deg(wP) \le d + w,$$

and so Theorem 3.7 yields the desired result.   □

Two variants of this construction of almost perfect sequences were introduced by Xing *et al.* [87]. For the first variant, let again $f \in F$ with $f \notin \mathbb{F}_q(t)$ and let $v = \nu_P(f)$ be arbitrary. Then the local expansion of f at P has the form

$$f = t^v \sum_{i=0}^{\infty} s_i t^i$$

with all $s_i \in \mathbb{F}_q$. From this expansion we read off the sequence S_2 of coefficients $s_0, s_1, \ldots$.

Theorem 3.10. *If $v = \nu_P(f)$ and the integer d is such that $d \geq \deg((f)_\infty)$, then the sequence S_2 constructed above is $(d + v - 1)$-perfect if $v > 0$ and $(d - v + 1)$-perfect if $v \leq 0$.*

Proof. First let $v > 0$. Then

$$t^{1-v} f = \sum_{i=1}^{\infty} s_{i-1} t^i.$$

Furthermore, by using $\nu_P(t) = 1$ and $\deg((t)_\infty) = 2$, we obtain

$$\deg((t^{1-v} f)_\infty) \leq \deg((f)_\infty) + v - 1 \leq d + v - 1.$$

Thus, by applying Theorem 3.7 with f replaced by $t^{1-v} f$, we get the first part of the theorem. The second part follows by adapting the argument in the proof of Theorem 3.7.   □

Example 3.11. Let $q = 3$, let F be the rational function field $\mathbb{F}_3(x)$, and let P be the zero of x. We choose $t = x^2 - x$ and $f = x$. Then at P we have the local expansion

$$f = -t + t^2 + t^3 - t^4 + t^5 + 0 \cdot t^6 + \cdots.$$

The sequence S_2 of coefficients $-1, 1, 1, -1, 1, 0, \ldots$ is perfect by Theorem 3.10.

The second construction variant from [87] is obtained as follows. Let $f \in F$ with $f \notin \mathbb{F}_q(t)$ and $\nu_P(f) \leq 0$. Put $w = -\nu_P(f) \geq 0$. Then the local expansion of f at P can be written in the form

$$f = \sum_{j=1}^{w} r_j t^{j-w-1} + \sum_{i=0}^{\infty} s_i t^i$$

with all $r_j \in \mathbb{F}_q$ and $s_i \in \mathbb{F}_q$. From this expansion we read off the sequence S_3 of coefficients $s_1, s_2, \ldots$. The following result is shown by the same method as Theorem 3.7.

Theorem 3.12. *If $\nu_P(f) \leq 0$ and the integer d is such that $d \geq \deg((f)_\infty)$, then the sequence S_3 constructed above is d-perfect.*

Further examples of almost perfect sequences that are obtained from the three theorems in this subsection can be found in Kohel, Ling and Xing [34], Xing [79], [81], and Xing and Niederreiter [86]. The paper of Kohel, Ling and Xing [34] also discusses the efficient computation of local expansions by means of an effective form of Hensel's lemma.

3.4   Generalisation to Multisequences

A multisequence is a parallel stream of finitely many sequences. We will denote an m-fold multisequence consisting of m parallel streams of sequences $S_1, \ldots, S_m$ over $\mathbb{F}_q$ by $\mathbf{S} = (S_1, \ldots, S_m)$. In the framework of linear complexity theory, the appropriate complexity measure for multisequences is obtained by looking at the linear recurrence relations that initial segments of $S_1, \ldots, S_m$ satisfy simultaneously.

Definition 3.13. Let n be a positive integer and let $\mathbf{S} = (S_1, \ldots, S_m)$ be an m-fold multisequence over $\mathbb{F}_q$. Then the *nth joint linear complexity* $L_n^{(m)}(\mathbf{S})$ of $\mathbf{S}$ is the least order of a linear recurrence relation over $\mathbb{F}_q$ that simultaneously generates the first n terms of each sequence S_j, $1 \leq j \leq m$. The sequence $L_1^{(m)}(\mathbf{S}), L_2^{(m)}(\mathbf{S}), \ldots$ is called the *joint linear complexity profile* of $\mathbf{S}$.

As in the case of single sequences, we again have $0 \leq L_n^{(m)}(\mathbf{S}) \leq n$ and $L_n^{(m)}(\mathbf{S}) \leq L_{n+1}^{(m)}(\mathbf{S})$. The analysis of the joint linear complexity profile is considerably more complicated than for the linear complexity profile. Note that for single sequences we have the limit relation (3.1) outside of a set of sequences S with μ_q^∞-measure 0. Let $\mu_{q,m}^\infty$ be the corresponding probability measure on the space of m-fold multisequences over $\mathbb{F}_q$. Then, according to a folklore conjecture (see e.g. [75] and [78]), we have

$$\lim_{n \to \infty} \frac{L_n^{(m)}(\mathbf{S})}{n} = \frac{m}{m+1} \qquad (3.3)$$

except on a set of m-fold multisequences $\mathbf{S}$ with $\mu_{q,m}^\infty$-measure 0. This conjecture was shown in [75] for $m = 2$. Recently, the conjecture was proved for all m by Niederreiter and Wang [42].

The limit relation (3.3) and Proposition 3.6 suggest the following extension of the definition of a d-perfect sequence.

Definition 3.14. Let d be a positive integer. Then an m-fold multisequence **S** over $\mathbb{F}_q$ is called *d-perfect* if

$$L_n^{(m)}(\mathbf{S}) \geq \frac{m(n+1) - d}{m + 1} \qquad \text{for all } n \geq 1.$$

A multisequence is called *almost perfect* if it is d-perfect for some d.

Xing [78] showed that if an m-fold multisequence over $\mathbb{F}_q$ is d-perfect, then d must be at least m. In the same paper, the following construction of almost perfect m-fold multisequences over $\mathbb{F}_q$ based on global function fields was introduced.

Let $F/\mathbb{F}_q$ be a global function field and let Q be a place of F of degree m. Let t be a local parameter at Q with $\deg((t)_\infty) = m + 1$. Since the residue class field F_Q of Q satisfies $[F_Q : \mathbb{F}_q] = m$, we can choose $x_1, \ldots, x_m \in F$ with $\nu_Q(x_j) \geq 0$ for $1 \leq j \leq m$ such that the residues $x_1(Q), \ldots, x_m(Q)$ form an $\mathbb{F}_q$-basis of F_Q. Finally, we choose $y \in F$ with $\nu_Q(y) \geq 0$ such that $y \notin \oplus_{i=1}^m \mathbb{F}_q(t)x_i$. The local expansion of y at Q has the form

$$y = \sum_{i=0}^{\infty} \left(\sum_{j=1}^{m} s_{i,j} x_j \right) t^i$$

with all $s_{i,j} \in \mathbb{F}_q$. Then for $1 \leq j \leq m$, let S_j be the sequence over $\mathbb{F}_q$ with terms $s_{1,j}, s_{2,j}, s_{3,j}, \ldots$.

Theorem 3.15. *Let* $\mathbf{S} = (S_1, \ldots, S_m)$ *be the m-fold multisequence over $\mathbb{F}_q$ obtained from the above construction. Then $\mathbf{S}$ is d-perfect with*

$$d = \deg((y)_\infty \vee (x_1)_\infty \vee \cdots \vee (x_m)_\infty).$$

Here $\vee$ denotes the maximum operation for divisors, i.e., if $D_1, D_2, \ldots, D_{m+1}$ are arbitrary divisors of F with

$$D_k = \sum_P n_P^{(k)} P \qquad \text{for } 1 \leq k \leq m + 1,$$

then

$$D_1 \vee D_2 \vee \cdots \vee D_{m+1} = \sum_P \max(n_P^{(1)}, n_P^{(2)}, \ldots, n_P^{(m+1)}) P.$$

The proof of Theorem 3.15 proceeds by generalising the argument in the proof of Theorem 3.7.

Xing [78] gave several examples for this construction. Further examples in which the m-fold multisequence is d-perfect with the least possible value $d = m$ were presented by Xing, Lam and Wei [85].

3.5 Sequences with Low Correlation and Large Linear Complexity

Global function fields can also be used for the construction of periodic sequences with low correlation and large linear complexity. A construction of this type was given by Xing, Kumar and Ding [83] for the binary case. We recall the definition of correlation for binary sequences.

Definition 3.16. Let $S = \{s_i\}_{i=1}^{\infty}$ and $T = \{t_i\}_{i=1}^{\infty}$ be two binary sequences of period r (it is allowed that S and T are the same). Then their *correlation* at shift $w \in \mathbb{Z}$ is given by

$$c_{S,T}(w) = \sum_{i=1}^{r} (-1)^{s_i + t_{i+w}}.$$

The construction in [83] proceeds as follows. Let q be a power of 2 and let $F/\mathbb{F}_q$ be a global function field. Choose a rational place P of F and an $\mathbb{F}_q$-automorphism σ of F. Note that $\sigma(P)$ is again a rational place of F. Put $P_i = \sigma^i(P)$ for all integers $i \geq 1$. For an element $z \in F$ with $\nu_{P_i}(z) \geq 0$ for all $i \geq 1$, define the binary sequence

$$S_z = \{\mathrm{Tr}(z(P_i))\}_{i=1}^{\infty}. \tag{3.4}$$

Here Tr denotes the trace function from $\mathbb{F}_q$ to $\mathbb{F}_2$ and $z(P_i)$ is, as usual, the residue of z in the residue class field of P_i.

If r is the least positive integer satisfying $\sigma^r(P) = P$, then it is clear that the sequence S_z in (3.4) is periodic with period r. As shown in [83], the following is a sufficient condition for r to be the least period of S_z: suppose that there exists a unique pole Q of z with $\nu_Q(z)$ odd, that $Q, \sigma(Q), \sigma^2(Q), \ldots, \sigma^{r-1}(Q)$ are distinct, and that $d = \deg((z)_\infty)$ satisfies

$$q + 1 + 2(2g(F) + 2d - 1)\sqrt{q} < 2r.$$

Under these conditions we also have the following lower bound on the linear complexity.

Theorem 3.17. *With the above notation and conditions, the linear complexity of the sequence S_z satisfies*

$$L(S_z) \geq \frac{2r - q - 1 - 2(2g(F) + d - 1)\sqrt{q}}{2d\sqrt{q}}.$$

Theorem 3.17 indicates that the linear complexity of S_z is large if the least period r is relatively large compared with q and $2g(F)\sqrt{q}$.

We recall that in the case of characteristic 2, an element $z \in F$ is called *nondegenerate* if it cannot be written in the form $a + h^2 + h$ for some $a \in \mathbb{F}_q$

and $h \in F$. A sufficient condition for z to be nondegenerate is that there exists a pole Q of z with $\nu_Q(z)$ odd. The following is an upper bound on the correlation of two sequences of the type (3.4).

Theorem 3.18. *Let $z_1, z_2 \in F$ with $\nu_{P_i}(z_j) \geq 0$ for $j = 1, 2$ and all $i \geq 1$. Suppose that $z_1 + \sigma^{-w}(z_2)$ is nondegenerate for some $w \in \mathbb{Z}$. Then we have*

$$|c_{S_{z_1}, S_{z_2}}(w)| \leq 2(2g(F) + b - 1)\sqrt{q} + |q + 1 - r| + 2(N(F) - r),$$

where b is the degree of the pole divisor of $z_1 + \sigma^{-w}(z_2)$.

Examples for this construction of sequences with low correlation and large linear complexity are given in the papers of Xing [79], [81] and Xing, Kumar and Ding [83].

References

[1] N. Alon, "Explicit construction of exponential sized families of k-independent sets", Discrete Math., Vol. 58, 191–193 (1986).

[2] M. Atici, S. S. Magliveras, D. R. Stinson and W. D. Wei, "Some recursive constructions for perfect hash families", J. Combinatorial Designs, Vol. 4, 353–363 (1996).

[3] J. Bierbrauer, "Universal hashing and geometric codes", Designs, Codes and Cryptography, Vol. 11, 207–221 (1997).

[4] J. Bierbrauer, T. Johansson, G. Kabatianskii and B. Smeets, "On families of hash functions via geometric codes and concatenation", *Advances in Cryptology – CRYPTO '93*, LNCS, Vol. 773, 331–342 (1994).

[5] S. R. Blackburn, "Combinatorics and threshold cryptology", *Combinatorial Designs and Their Applications*, Chapman and Hall/CRC Research Notes in Mathematics, CRC Press, London, 49–70 (1999).

[6] S. R. Blackburn, "Perfect hash families: probabilistic methods and explicit constructions", J. Combinatorial Theory Series A, Vol. 92, 54–60 (2000).

[7] S. R. Blackburn, "Frameproof codes", SIAM J. Discrete Math., Vol. 16, 499–510 (2003).

[8] S. R. Blackburn, M. Burmester, Y. Desmedt and P. R. Wild, "Efficient multiplicative sharing schemes", *Advances in Cryptology – EUROCRYPT '96*, LNCS, Vol. 1070, 107–118 (1996).

[9] S. R. Blackburn and P. R. Wild, "Optimal linear perfect hash families", J. Combinatorial Theory Series A, Vol. 83, 233–250 (1998).

[10] D. Boneh and J. Shaw, "Collision-secure fingerprinting for digital data", IEEE Trans. Inform. Theory, Vol. 44, 1897–1905 (1998).

[11] E. F. Brickell, "A problem in broadcast encryption", 5th Vermont Summer Workshop on Combinatorics and Graph Theory, June 1991.

[12] J. L. Carter and M. N. Wegman, "Universal classes of hash functions", J. Computer and System Sciences, Vol. 18, 143–154 (1979).

[13] B. Chor, A. Fiat and M. Naor, "Tracing traitors", *Advances in Cryptology – CRYPTO '94*, LNCS, Vol. 839, 257–270 (1994).

[14] G. Cohen and S. Encheva, "Efficient constructions of frameproof codes", Electronics Letters, Vol. 36, 1840–1842 (2000).

[15] Z. J. Czech, G. Havas and B. S. Majewski, "Perfect hashing", Theoretical Computer Science, Vol. 182, 1–143 (1997).

[16] Y. Desmedt, "Threshold cryptography", European Trans. on Telecommunications, Vol. 5(4), 449–457 (1994).

[17] Y. Desmedt, R. Safavi-Naini, H. Wang, L. M. Batten, C. Charnes and J. Pieprzyk, "Broadcast anti-jamming systems", Computer Networks, Vol. 35 (2-3), 223–236 (2001).

[18] A.G. Dyachkov and V.V. Rykov, "Bounds on the length of disjunctive codes" (in Russian), Problemy Peredachi Informatsii, Vol. 18, 7–13 (1982).

[19] P. Erdös, P. Frankl and Z. Füredi, "Families of finite sets in which no set is covered by the union of r others", Israel J. Math., Vol. 51, 79–89 (1985).

[20] A. Fiat and M. Naor, "Broadcast encryption", *Advances in Cryptology – CRYPTO '93*, LNCS, Vol. 773, 480–491 (1994).

[21] A. Fiat and T. Tassa, "Dynamic traitor tracing", *Advances in Cryptology – CRYPTO '99*, LNCS, Vol. 1666, 354–371 (1999).

[22] M. L. Fredman and J. Komlós, "On the size of separating systems and families of perfect hash functions", SIAM J. Alg. Discrete Methods, Vol. 5, 61–68 (1984).

[23] Z. Füredi, "On r-cover-free families", J. Combinatorial Theory Series A, Vol. 73, 172–173 (1996).

[24] A. Garcia and H. Stichtenoth, "A tower of Artin-Schreier extensions of function fields attaining the Drinfeld-Vladut bound", Invent. Math., Vol. 121, 211–222 (1995).

[25] A. Garcia and H. Stichtenoth, "On the asymptotic behaviour of some towers of function fields over finite fields", J. Number Theory, Vol. 61, 248–273 (1996).

[26] A. Garcia, H. Stichtenoth and C. P. Xing, "On subfields of the Hermitian function field", Compositio Math., Vol. 120, 137–170 (2000).

[27] E. N. Gilbert, F. J. MacWilliams and N. J. A. Sloane, "Codes which detect deception", The Bell System Technical Journal, Vol. 33 (3), 405–424 (1974).

[28] R. Hartshorne, *Algebraic Geometry*, Springer, New York, 1977.

[29] T. Helleseth and T. Johansson, "Universal hash functions from exponential sums over finite fields and Galois rings", *Advances in Cryptology – CRYPTO '96*, LNCS, Vol. 1109, 31–44 (1996).

[30] Y. Ihara, "Some remarks on the number of rational points of algebraic curves over finite fields", J. Fac. Sci. Univ. Tokyo Sect. IA Math., Vol. 28, 721–724 (1981).

[31] T. Johansson, *Contributions to unconditionally secure authentication*, Ph.D. thesis, Lund, 1994.

[32] G. Kabatianskii, B. Smeets and T. Johansson, "On the cardinality of systematic authentication codes via error-correcting codes", IEEE Trans. Inform. Theory, Vol. 42, 566–578 (1996).

[33] W. H. Kautz and R. C. Singleton, "Nonrandom binary superimposed codes", IEEE Trans. Inform. Theory, Vol. 10, 363–377 (1964).

[34] D. Kohel, S. Ling and C. P. Xing, "Explicit sequence expansions", *Sequences and Their Applications* (C. S. Ding, T. Helleseth and H. Niederreiter, eds.), Springer, London, 308–317 (1999).

[35] R. Kumar, S. Rajagopalan and A. Sahai, "Coding constructions for blacklisting problems without computational assumptions", *Advances in Cryptology – CRYPTO '99*, LNCS, Vol. 1666, 609–623 (1999).

[36] Yu. I. Manin, "What is the maximum number of points on a curve over $\mathbf{F}_2$?", J. Fac. Sci. Univ. Tokyo Sect. IA Math., Vol. 28, 715–720 (1981).

[37] K. Martin, J. Pieprzyk, R. Safavi-Naini, H. Wang and P. Wild, "Threshold MACs", *5th International Conference on Information Security and Cryptology (ICISC '02)*, LNCS, Vol. 2587, 237–252 (2003).

[38] K. Martin, R. Safavi-Naini, H. Wang and P. Wild, "Distributing the encryption and decryption of a block cipher", Designs, Codes and Cryptography, Vol. 36, 263–287 (2005).

[39] K. Mehlhorn, *Data Structures and Algorithms*, Volume 1, Springer, Berlin, 1984.

[40] C. J. Mitchell and F. C. Piper, "Key storage in secure networks", Discrete Applied Math., Vol. 21, 215–228 (1988).

[41] D. Mumford, *Abelian Varieties*, Oxford University Press, Oxford, 1970.

[42] H. Niederreiter and L.-P. Wang, "Proof of a conjecture on the joint linear complexity profile of multisequences", *Progress in Cryptology – INDOCRYPT 2005*, LNCS, Vol. 3797, 13–22 (2005).

[43] H. Niederreiter and C. P. Xing, "Explicit global function fields over the binary field with many rational places", Acta Arith., Vol. 75, 383–396 (1996).

[44] H. Niederreiter and C. P. Xing, "Low-discrepancy sequences and global function fields with many rational places", Finite Fields Appl., Vol. 2, 241–273 (1996).

[45] H. Niederreiter and C. P. Xing, "Towers of global function fields with asymptotically many rational places and an improvement on the Gilbert-Varshamov bound", Math. Nachr., Vol. 195, 171–186 (1998).

[46] H. Niederreiter and C. P. Xing, *Rational Points on Curves over Finite Fields: Theory and Applications*, Cambridge University Press, Cambridge, 2001.

[47] H. Niederreiter and C. P. Xing, "Constructions of digital nets", Acta Arith., Vol. 102, 189–197 (2002).

[48] J. Pieprzyk, H. Wang and C. P. Xing, "Multiple-time signature schemes secure against adaptive chosen message attacks", *10th Workshop on Selected Areas in Cryptography (SAC '03)*, LNCS, Vol. 3006, 88–100 (2004).

[49] R. A. Rueppel, Stream ciphers, *Contemporary Cryptology: The Science of Information Integrity* (G. J. Simmons, ed.), IEEE Press, New York, 65–134 (1992).

[50] M. Ruszinkó, On the upper bound of the size of the r-cover-free families, J. Combinatorial Theory Series A, Vol. 66, 302–310 (1994).

[51] R. Safavi-Naini and H. Wang, "New results on multireceiver authentication codes", *Advances in Cryptology – EUROCRYPT '98*, LNCS, Vol. 1403, 527–541 (1998).

[52] R. Safavi-Naini and H. Wang, "New constructions of secure multicast re-keying schemes using perfect hash families", *7th ACM Conference on Computer and Communication Security*, ACM Press, 228–234 (2000).

[53] R. Safavi-Naini and H. Wang, "Efficient authentication for group communication", Theoretical Computer Science, Vol. 269, 1–21 (2001).

[54] R. Schoof, "Algebraic curves over $\mathbf{F}_2$ with many rational points", J. Number Theory, Vol. 41, 6–14 (1992).

[55] J. P. Serre, "Sur le nombre des points rationnels d'une courbe algébrique sur un corps fini", C. R. Acad. Sci. Paris Sér. I Math., Vol. 296, 397–402 (1983).

[56] J. P. Serre, "Nombres de points des courbes algébriques sur $\mathbf{F}_q$", Sém. Théorie des Nombres 1982-1983, Exp. 22, Université de Bordeaux I, Talence, 1983.

[57] J. P. Serre, Rational Points on Curves over Finite Fields, Lecture Notes, Harvard University, 1985.

[58] A. Shamir, "How to share a secret", Communications of the ACM, Vol. 22, 612–613 (1979).

[59] G. J. Simmons, "Authentication theory/coding theory", Advances in Cryptology – CRYPTO '84, LNCS, Vol. 196, 411–431 (1984).

[60] G. J. Simmons, "A survey of information authentication", Contemporary Cryptology: The Science of Information Integrity (G. J. Simmons, ed.), IEEE Press, New York, 379–419 (1992).

[61] J. N. Staddon, D. R. Stinson and R. Wei, "Combinatorial properties of frameproof and traceability codes", IEEE Trans. Inform. Theory, Vol. 47, 1042–1049 (2001).

[62] H. Stichtenoth, Algebraic Function Fields and Codes, Springer, Berlin, 1993.

[63] D. R. Stinson, "Combinatorial characterization of authentication codes", Designs, Codes and Cryptography, Vol. 2, 175–187 (1992).

[64] D. R. Stinson, "Universal hashing and authentication codes", Designs, Codes and Cryptography, Vol. 4, 369–380 (1994); also Advances in Cryptology – CRYPTO '91, LNCS, Vol. 576, 74–85 (1992).

[65] D. R. Stinson, "On the connection between universal hashing, combinatorial designs and error-correcting codes", Congressus Numerantium, Vol. 114, 7–27 (1996).

[66] D. R. Stinson, "On some methods for unconditionally secure key distribution and broadcast encryption", Designs, Codes and Cryptography, Vol. 12, 215–243 (1997).

[67] D. R. Stinson, T. van Trung and R. Wei, "Secure frameproof codes, key distribution patterns, group testing algorithms and related structures", J. Statist. Plan. Infer., Vol. 86, 595–617 (2000).

[68] D. R. Stinson and R. Wei, "Combinatorial properties and constructions of traceability schemes and frameproof codes", SIAM J. Discrete Math., Vol. 11, 41–53 (1998).

[69] D. R. Stinson, R. Wei and L. Zhu, "New constructions for perfect hash families and related structures using combinatorial designs and codes", J. Combinatorial Designs, Vol. 8, 189–200 (2000).

[70] D. R. Stinson, R. Wei and L. Zhu. "Some new bounds for cover-free families", J. Combinatorial Theory Series A, Vol. 90, 224–234 (2000).

[71] M. A. Tsfasman, S. G. Vlăduţ and T. Zink, "Modular curves, Shimura curves, and Goppa codes, better than Varshamov-Gilbert bound", Math. Nachr., Vol. 109, 21–28 (1982).

[72] G. van der Geer and M. van der Vlugt, "Tables of curves with many points", Math. Comp., Vol. 69, 797–810 (2000).

[73] S. G. Vlăduţ and V. G. Drinfeld, "Number of points of an algebraic curve", Funct. Anal. Appl., Vol. 17, 53–54 (1983).

[74] H. Wang and C. P. Xing, "Explicit constructions of perfect hash families from algebraic curves over finite fields", J. Combinatorial Theory Series A, Vol. 93, 112–124 (2001).

[75] L.-P. Wang and H. Niederreiter, "Enumeration results on the joint linear complexity of multisequences", Finite Fields Appl., to appear.

[76] M. N. Wegman and J. L. Carter, "New hash functions and their use in authentication and set equality", J. Computer and System Sciences, Vol. 22, 265–279 (1981).

[77] R. Wei, "On cover-free families", Discrete Math., to appear.

[78] C. P. Xing, "Multi-sequences with almost perfect linear complexity profile and function fields over finite fields", J. Complexity, Vol. 16, 661–675 (2000).

[79] C. P. Xing, "Applications of algebraic curves to constructions of sequences", *Cryptography and Computational Number Theory* (K.-Y. Lam *et al.*, eds.), Birkhäuser, Basel, 137–146 (2001).

[80] C. P. Xing, "Algebraic-geometry codes with asymptotic parameters better than the Gilbert-Varshamov and the Tsfasman-Vlăduţ-Zink bounds", IEEE Trans. Inform. Theory, Vol. 47, 347–352 (2001).

[81] C. P. Xing, "Constructions of sequences from algebraic curves over finite fields", *Sequences and Their Applications – SETA '01* (T. Helleseth, P. V. Kumar and K. Yang, eds.), Springer, London, 88–100 (2002).

[82] C. P. Xing, "Asymptotic bounds on frameproof codes", IEEE Trans. Inform. Theory, Vol. 48, 2991–2995 (2002).

[83] C. P. Xing, P. V. Kumar and C. S. Ding, "Low-correlation, large linear span sequences from function fields", IEEE Trans. Inform. Theory, Vol. 49, 1439–1446 (2003).

[84] C. P. Xing and K. Y. Lam, "Sequences with almost perfect linear complexity profiles and curves over finite fields", IEEE Trans. Inform. Theory, Vol. 45, 1267–1270 (1999).

[85] C. P. Xing, K. Y. Lam and Z. H. Wei, "A class of explicit perfect multi-sequences", *Advances in Cryptology – ASIACRYPT '99* (K. Y. Lam, E. Okamoto and C. P. Xing, eds.), LNCS, Vol. 1716, 299–305 (1999).

[86] C. P. Xing and H. Niederreiter, "Applications of algebraic curves to constructions of codes and almost perfect sequences", *Finite Fields and Applications* (D. Jungnickel and H. Niederreiter, eds.), Springer, Berlin, 475–489 (2001).

[87] C. P. Xing, H. Niederreiter, K. Y. Lam and C. S. Ding, "Constructions of sequences with almost perfect linear complexity profile from curves over finite fields", Finite Fields Appl., Vol. 5, 301–313 (1999).

[88] C. P. Xing, H. Wang and K. Y. Lam, "Constructions of authentication codes from algebraic curves over finite fields", IEEE Trans. Inform. Theory, Vol. 46, 886–892 (2000).

Chapter 3

ARTIN-SCHREIER EXTENSIONS AND THEIR APPLICATIONS

Cem Güneri and Ferruh Özbudak

1.   Introduction

A Galois extension E/F of fields is called a *cyclic extension* if the Galois group is cyclic. Assume that $p > 0$ is the characteristic of our fields and n is the degree of the field extension E/F. If n is relatively prime to p, and there is a primitive n^{th} root of unity in F, then E/F is a *Kummer extension*, i.e. $E = F(y)$ with $y^n \in F$. If $n = p$, then E/F is an *Artin-Schreier extension*, i.e. $E = F(y)$ with $y^p - y \in F$. Finally, if $n = p^a$ for $a > 1$, then the extension E/F can be described in terms of *Witt vectors*. For these facts, see [34, Section VI.7].

In this survey, we are interested in Artin-Schreier extensions of function fields and their generalizations. Namely, we will have a Galois extension E/F of function fields of degree $q = p^e$ ($e \geq 1$) whose Galois group is isomorphic to the direct sum of e copies of $\mathbb{Z}/p\mathbb{Z}$. When $e = 1$ this is an Artin-Schreier extension as mentioned above. When $e > 1$ such an extension is called an *elementary abelian p-extension* due to the structure of its Galois group. We will in general refer to all such extensions as *Artin-Schreier type* extensions.

Besides introducing some basic properties of Artin-Schreier type extensions, the second purpose of this chapter is to explain some applications of such function field extensions to coding theory. We remind the reader that since the introduction of algebraic geometry codes by Goppa in [23], interaction between coding theory and algebraic function fields (curves) over finite fields has been explored by many researchers, both mathematicians and engineers. Our focus will be on another well-established relation between these two topics, which

A. Garcia and H. Stichtenoth (eds.), Topics in Geometry, Coding Theory and Cryptography, 105–133.

arises via the trace map between finite fields. By Hilbert's Theorem 90 and Delsarte's theorem, weights (and higher weights) of the so-called trace codes are intimately related to Artin-Schreier function fields.

Let us give a short summary. In Section 2, we introduce Artin-Schreier type extensions and specifically address the genus and irreducibility issues. In the same section, a relation between the number of solutions of defining equations of Artin-Schreier type extensions and certain character sums is established. It is shown that the Hasse-Weil bound for Artin-Schreier function fields can be used to obtain the Weil bound for character sums.

Section 3 is devoted to applications of Artin-Schreier type extensions to weight analysis of cyclic codes. After reviewing basic material on cyclic codes, we prove their trace representation and state Wolfmann's bound on the weights of cyclic codes using the Hasse-Weil theorem. We also illustrate an extension of Wolfmann's bound to a larger class of cyclic codes and describe how similar methods can be adapted to multidimensional cyclic codes.

In general, a code over $\mathbb{F}_q$ which can be described as the image under the trace map of another code over an extension of $\mathbb{F}_q$ is called a *trace code*. In this sense, cyclic codes are also trace codes. Section 4 deals with trace codes and the application of Artin-Schreier type extensions to their "ordinary" weights and higher weights. We explore this relation when $q = p$ is a prime and mention the complications in the case $q = p^e$ with $e > 1$.

In the last section we summarize some of the developments in the explicit construction of function fields over finite fields with many rational places. Artin-Schreier type extensions yield many "good" examples in this respect. We note that although the topic addressed in this section is purely mathematical, the interest in recent times on this topic was motivated by applications to coding theory. We finish by recalling a theorem of Frey-Perret-Stichtenoth which states that although one can find a lot of examples of Artin-Schreier type function fields with many rational places, this class is "asymptotically bad".

As is the case with most publications in this area, one has to make a choice between the language of curves and function fields. We will mostly use terminology and notation from function fields, for which a reader can consult the appendix of this volume or [50]. Yet, we remind the fact that the two approaches are equivalent and we will feel free to use the geometric language as well.

Unless otherwise stated the following notation will be valid throughout this chapter.

- $\mathbb{F}_q$ is a finite field with $q = p^e$ elements, where p is a prime number.

- F is a function field over $\mathbb{F}_q$.

- $m \geq 2$ is an integer.

- $\mathrm{Tr}_{\mathbb{F}_{q^m}/\mathbb{F}_q}$ and $\mathrm{Tr}_{\mathbb{F}_{q^m}/\mathbb{F}_p}$ are the trace maps from $\mathbb{F}_{q^m}$ onto $\mathbb{F}_q$ and $\mathbb{F}_p$.

2.   Artin-Schreier Extensions

In this section we introduce a class of abelian extensions of algebraic function fields in positive characteristic. Namely, given a function field F of characteristic $p > 0$, we are interested in Galois extensions E of F such that the Galois group $Gal(E/F)$ is either $\mathbb{Z}/p\mathbb{Z}$ or a direct sum of a finite number of copies of $\mathbb{Z}/p\mathbb{Z}$. Although we will refer to this kind of extensions as Artin-Schreier (or Artin-Schreier type), due to the structure of Galois group, they are also called elementary abelian p-extensions. The rest of the chapter will be mainly based on the results stated in this section. At the end of the section, we discuss a relation between the number of rational places of Artin-Schreier function fields and certain character sums.

We start with the simplest case:

Definition 2.1. For $f \in F$, assume that $T^p - T - f \in F[T]$ is irreducible. Let $E = F(y)$ with $y^p - y = f$. Then E/F is called an *Artin-Schreier* extension.

An Artin-Schreier extension E/F is a Galois extension with Galois group $Gal(E/F) = \{\sigma_i : y \mapsto y+i,\ i \in \mathbb{F}_p\}$, which is isomorphic to $\mathbb{Z}/p\mathbb{Z}$. A place P of F is either unramified or totally ramified in E/F. In particular if $v_P(f) \geq 0$, then P is unramified. If there exists a ramified place in the extension, then E and F have the same constant field (see [50, Proposition III.7.8] for the proofs of these assertions). Let us also remark that in characteristic $p > 0$, any Galois extension of degree p is an Artin-Schreier extension ([34, Theorem 6.4]).

The genus of E can be computed using the genus of F and Riemann-Hurwitz formula ([50, Proposition III.4.12]). In particular if $F = \mathbb{F}_q(x)$ is the rational function field and f is a polynomial in $\mathbb{F}_q[x]$ with $\gcd(\deg f, p) = 1$, then $T^p - T - f$ defines an Artin-Schreier extension over F. In this extension the place at infinity is the only ramified place and the genus $g(E)$ of E is

$$g(E) = \frac{(p-1)(\deg f - 1)}{2}. \tag{2.1}$$

Next, we introduce a class of elementary abelian extensions, which includes the class of Artin-Schreier extensions.

Definition 2.2. Let F be a function field with $\mathbb{F}_q \subset F$. Assume that the polynomial $T^q - T - f \in F[T]$ is irreducible. Let $E = F(y)$ with $y^q - y = f$. We call E/F an *Artin-Schreier type* extension.

The following result, and more on Artin-Schreier type extensions, can be found in [11].

Theorem 2.3. *Let F be a function field with $\mathbb{F}_q \subset F$.*

(i) If E/F is an elementary abelian extension of degree q, then there exists an element $y \in E$ with $E = F(y)$ whose minimal polynomial over F has the form $T^q - T - f$ for some $f \in F$.

(ii) Conversely, if $T^q - T - f \in F[T]$ is irreducible over F, then the extension $F(y)/F$ with $y^q - y = f$ is an elementary abelian extension of degree q. There are $(q-1)/(p-1)$ intermediate fields $F \subset E_a \subset F(y)$ with $[E_a : F] = p$. These are defined by $E_a = F(y_a)$, where $a \in \mathbb{F}_q^*$ and

$$y_a = (ay)^{p^{e-1}} + (ay)^{p^{e-2}} + \cdots + (ay)^p + (ay).$$

The element y_a satisfies the equation $y_a^p - y_a = az$.

(iii) $T^q - T - z \in F[T]$ is irreducible over F if and only if $T^p - T - az$ is irreducible over F for all $a \in \mathbb{F}_q^*$.

(iv) Assume that F is the rational function field whose constant field is finite and contains $\mathbb{F}_q$. If E is an elementary abelian extension of F of degree q with the same constant field, and $E_1, \ldots, E_v$ denote the intermediate fields $F \subset E_i \subset E$ with $[E_i : F] = p$ for all $i = 1, \ldots, v = (q-1)/(p-1)$, then the genus $g(E)$ of E is

$$g(E) = \sum_{i=1}^{v} g(E_i).$$

Example 2.4. Assume that $F = \mathbb{F}_{q^m}(x)$, $f \in \mathbb{F}_{q^m}[x]$ and $\gcd(\deg f, p) = 1$. Then $T^q - T - f \in F[T]$ is irreducible. Let $E = F(y)$ with $y^q - y = f$. Using Theorem 2.3 (iv) and Equation (2.1), for the genus $g(E)$ of E we obtain that

$$g(E) = \frac{(q-1)(\deg f - 1)}{2}. \tag{2.2}$$

For applications and also due to its theoretical appeal, the number of rational places of Artin-Schreier extensions is of great interest. Although we have a separate section (Section 5) on this issue, we start the discussion of the topic here. This will also enable us to explain a relation with character sums.

Recall that *Hilbert's Theorem 90* states that for $m \geq 1$ and $\alpha \in \mathbb{F}_{q^m}$,

$$\mathrm{Tr}_{\mathbb{F}_{q^m}/\mathbb{F}_q}(\alpha) = 0 \Longleftrightarrow \alpha = \beta^q - \beta \text{ for some } \beta \in \mathbb{F}_{q^m}. \tag{2.3}$$

Note that if there exists an element $\beta \in \mathbb{F}_{q^m}$ with $\beta^q - \beta = \alpha$, then for any $c \in \mathbb{F}_q$, the element $\beta + c$ also satisfies the same equation. Combining this observation with the ramification structure of Artin-Schreier type extensions and Kummer's theorem ([50, Theorem III.3.7]), we obtain the following:

Proposition 2.5. *Let F be a function field whose constant field is $\mathbb{F}_{q^m}$. Assume that $T^q - T - f \in F[T]$ is irreducible and $E = F(y)$ with $y^q - y = f$. For any rational place P of F with $v_P(f) \geq 0$ we have:*

i) if $\mathrm{Tr}_{\mathbb{F}_{q^m}/\mathbb{F}_q}(f(P)) = 0$, then there are q rational places of E over P,

ii) if $\mathrm{Tr}_{\mathbb{F}_{q^m}/\mathbb{F}_q}(f(P)) \neq 0$, then there is no rational place of E over P, where $f(P) \in \mathbb{F}_{q^m}$ denotes the evaluation (residue class) of f at P.

Another consequence of Hilbert's Theorem 90 is the following.

Proposition 2.6. *Let* $f(x) \in \mathbb{F}_{q^m}[x]$ *be a polynomial. Then, the number of solutions of the equation* $y^q - y = f(x)$ *in* $\mathbb{F}_{q^m} \times \mathbb{F}_{q^m}$ *is divisible by* q *and it is bounded by* q^{m+1}.

We refer the reader to [24, Theorem 2.5] for a characterization of the polynomials $f(x)$ that yield the maximum possible number of solutions in this Proposition.

Assume that $f(x) \in \mathbb{F}_{q^m}[x]$ and $\gcd(\deg f, p) = 1$. Let N_f denote the number solutions of the equation

$$y^q - y = f(x),$$

where $(x, y) \in \mathbb{F}_{q^m} \times \mathbb{F}_{q^m}$. Let $E = \mathbb{F}_{q^m}(x, y)$ be the function field defined by this equation and $N(E)$ denote the number of rational places of E. Then E has just one rational place at infinity and N_f affine rational places, i.e.

$$N(E) = 1 + N_f. \tag{2.4}$$

By the Hasse-Weil bound ([50, Theorem V.2.3]) and (2.2) we obtain that

$$|N(E) - (q^m + 1)| \le 2g(E)q^{m/2} = (q - 1)(\deg f - 1)q^{m/2}. \tag{2.5}$$

In the rest of this section we want to show that the bound (2.5) for the rational places of certain Artin-Schreier type function fields can be used to prove Weil's Theorem on additive character sums. Before stating this theorem we need some preparation.

Definition 2.7. An *additive character* ψ of $\mathbb{F}_{q^m}$ is a homomorphism from the additive group of $\mathbb{F}_{q^m}$ into the multiplicative group of $\mathbb{C}$, i.e.

$$\psi(x + y) = \psi(x)\psi(y) \quad \text{for } x, y \in \mathbb{F}_{q^m}.$$

For $a \in \mathbb{F}_{q^m}$, the map $\psi_a : \mathbb{F}_{q^m} \to \mathbb{C}^*$ defined by

$$\psi_a(x) = e^{\frac{2\pi i}{p} \operatorname{Tr}_{\mathbb{F}_{q^m}/\mathbb{F}_p}(ax)},$$

where $\mathbb{F}_p$ is identified with the set $\{0, 1, \ldots, p - 1\}$, is an additive character. Moreover any additive character of $\mathbb{F}_{q^m}$ is of this form for some $a \in \mathbb{F}_{q^m}$ (cf. [86, Theorem 5.7]). The character ψ_0 is called the *trivial additive character* of $\mathbb{F}_{q^m}$. Note that the image of an additive character lies in the unit circle $|z| = 1$ in $\mathbb{C}$.

We recall Weil's Theorem on additive character sums (cf. [86, Theorem 5.38]).

Theorem 2.8. *If $f(x) \in \mathbb{F}_{q^m}[x]$ is a polynomial with $\gcd(\deg f, p) = 1$ and ψ is a nontrivial additive character of $\mathbb{F}_{q^m}$, then we have*

$$\left| \sum_{x \in \mathbb{F}_{q^m}} \psi(f(x)) \right| \leq (\deg f - 1) q^{m/2}.$$

Remark 2.9. It is possible to prove Theorem 2.8 using only "elementary methods", the so-called *Stepanov-Schmidt method* instead of the Hasse-Weil bound. We refer the reader to [86, Chapter 6] for details.

For a nontrivial additive character ψ of $\mathbb{F}_{q^m}$ and $s \geq 1$, let $\psi^{(s)}$ be the additive character of $\mathbb{F}_{q^{ms}}$ defined by

$$\psi^{(s)}(x) = \psi\left(\mathrm{Tr}_{\mathbb{F}_{q^{ms}}/\mathbb{F}_{q^m}}(x) \right).$$

We will need the following results for the proof of Theorem 2.8. For their proofs, we refer to [86, Theorem 5.36, Lemma 6.55].

Theorem 2.10. *For $f \in \mathbb{F}_{q^m}[x]$ with $\deg f \geq 2$, $\gcd(\deg f, p) = 1$ and a nontrivial additive character ψ of $\mathbb{F}_{q^m}$, there exist nonzero complex numbers $w_1, \ldots, w_{\deg f - 1}$ such that for each $s \geq 1$ we have*

$$\sum_{x \in \mathbb{F}_{q^{ms}}} \psi^{(s)}(f(x)) = -w_1^s - \cdots - w_{\deg f - 1}^s.$$

Lemma 2.11. *Let $w_1, \ldots, w_n$ be nonzero complex numbers and $B > 0, C > 0$ be constants. Assume that for each $s \geq 1$ we have*

$$|w_1^s + \cdots + w_n^s| \leq CB^s.$$

Then $|w_i| \leq B$ for each $1 \leq i \leq n$.

We are ready to prove Theorem 2.8. For $s \geq 1$ and $a \in \mathbb{F}_{q^m}$, let $N_s(a)$ denote the number of $x \in \mathbb{F}_{q^{ms}}$ such that $\mathrm{Tr}_{\mathbb{F}_{q^{ms}}/\mathbb{F}_{q^m}}(f(x)) = a$. Using Proposition 2.5 we note that the number of rational places of the function field $\mathbb{F}_{q^{ms}}(x, y)$, where

$$y^{q^m} - y = f(x) - a, \tag{2.6}$$

is $q^m N_s(a) + 1$. Using the Hasse-Weil bound (2.5), we have

$$|N_s(a) - q^{m(s-1)}| \leq \frac{q^m - 1}{q^m}(\deg f - 1)q^{ms/2}. \tag{2.7}$$

Let $R_s(a) = N_s(a) - q^{m(s-1)}$. For $s \geq 1$ we have

$$
\begin{aligned}
\sum_{x \in \mathbb{F}_{q^{ms}}} \psi^{(s)}(f(x)) &= \sum_{a \in \mathbb{F}_{q^m}} \psi(N_s(a)a) \\
&= \sum_{a \in \mathbb{F}_{q^m}} N_s(a)\psi(a) \\
&= \sum_{a \in \mathbb{F}_{q^m}} \left(R_s(a) + q^{m(s-1)} \right) \psi(a).
\end{aligned}
$$

Since ψ is a nontrivial caharacter of $\mathbb{F}_{q^m}$, there exists $b \in \mathbb{F}_{q^m}$ with $\psi(b) \neq 0$. Then

$$
\psi(b) \sum_{a \in \mathbb{F}_{q^m}} \psi(a) = \sum_{a \in \mathbb{F}_{q^m}} \psi(a + b) = \sum_{a \in \mathbb{F}_{q^m}} \psi(a),
$$

and hence $\sum_{a \in \mathbb{F}_{q^m}} \psi(a) = 0$. Then for $s \geq 1$,

$$
\sum_{x \in \mathbb{F}_{q^{ms}}} \psi^{(s)}(f(x)) = \sum_{a \in \mathbb{F}_{q^m}} R_s(a)\psi(a),
$$

and using (2.7) we get

$$
\left| \sum_{x \in \mathbb{F}_{q^{ms}}} \psi^{(s)}(f(x)) \right| \leq \sum_{a \in \mathbb{F}_{q^m}} |R_s(a)| \leq (q^m - 1)(\deg f - 1)q^{ms/2}. \quad (2.8)
$$

Assume that $w_1, \ldots, w_{\deg f - 1}$ are nonzero complex numbers such that

$$
\left| \sum_{x \in \mathbb{F}_{q^{ms}}} \psi^{(s)}(f(x)) \right| = \left| w_1^s + \cdots w_{\deg f - 1}^s \right|
$$

for all $s \geq 1$ (cf. Theorem 2.10). Then by (2.8) and Lemma 2.11 we have $|w_i| \leq q^{m/2}$ for each $i = 1, \ldots, \deg f - 1$. Hence we have

$$
\left| \sum_{x \in \mathbb{F}_{q^m}} \psi(f(x)) \right| = \left| \sum_{i=1}^{\deg f - 1} w_i \right| \leq \sum_{i=1}^{\deg f - 1} |w_i| \leq (\deg f - 1)q^{m/2},
$$

which proves Theorem 2.8.

3. Cyclic Codes and Their Weights

Determining the minimum distance of a code is one of the most fundamental problems in coding theory. This is a difficult problem in general. Therefore, finding effective bounds for the minimum distance of a code is of great interest. The main goal in this section is to obtain a general bound on the minimum distance of a large class of cyclic codes (due to J. Wolfmann, see [61]), which is, in particular, valid for any p-ary cyclic code (cf. Remark 3.6). Our tools will be Artin-Schreier type extensions of algebraic function fields and the Hasse-Weil bound.

For more information on cyclic codes, and algebraic coding theory in general, we refer to [36, 40, 43]. A good survey of cyclic codes and some open problems on the subject are provided in [2].

We start with defining basic notions about codes. A *q-ary code of length* n is simply a subset C of $\mathbb{F}_q^n$. If C is a linear subspace of $\mathbb{F}_q^n$, then we call it a *linear code*. In this Chapter, the term "code" is only used for linear codes. If C is a k-dimensional subspace of $\mathbb{F}_q^n$, then it is called a code of *length* n and *dimension* k. An element of a code is called a *codeword*. *The minimum distance of a code* is defined as the minimum nonzero codeword weight, where *the weight of a codeword* is the number of nonzero coordinates in it. A linear code with length n, dimension k, and minimum distance d is called an $[n, k, d]$ code. These three parameters of a code are related via the *Singleton bound* ([36, page 67]), which states that $k + d \leq n + 1$. If the parameters of a code yield equality in the Singleton bound, then such a code is called *maximum distance separable* (MDS).

Definition 3.1. A linear code C with the property that if $(c_0, c_1, \ldots, c_{n-1})$ is in C, then $(c_{n-1}, c_0, \ldots, c_{n-2})$ is also in C is called a *cyclic* code.

The set of n-tuples that are orthogonal to all members of the code C, with respect to the usual inner product on $\mathbb{F}_q^n$, is called *the dual code* and is denoted by $C^\perp$. Observe that the dual of a cyclic code is also cyclic.

One of the most important features of cyclic codes is that they can be represented as ideals in certain rings. For this, observe the $\mathbb{F}_q$-isomorphism between $\mathbb{F}_q^n$ and $\mathbb{F}_q[t]/(t^n - 1)$:

$$(a_0, a_1, \ldots, a_{n-1}) \in \mathbb{F}_q^n \longleftrightarrow a(t) = \sum_{i=0}^{n-1} a_i t^i \in \mathbb{F}_q[t]/(t^n - 1) \qquad (3.1)$$

Under this identification a codeword $c \in C$ can now be viewed as a polynomial $c(t)$ and in this way we can think of a cyclic code as a subset of $\mathbb{F}_q[t]/(t^n - 1)$. It is easy to see that a linear code C in $\mathbb{F}_q^n$ is cyclic if and only if C is an ideal in $\mathbb{F}_q[t]/(t^n - 1)$.

Since a cyclic code is an ideal in the principal ideal ring $\mathbb{F}_q[t]/(t^n - 1)$, it is generated by a unique monic polynomial $g(t)$ of lowest degree in $\mathbb{F}_q[t]$ which divides $t^n - 1$. This polynomial is called *the generator polynomial* of the cyclic code. Besides other uses, the generator polynomial gives us a simple way to compute the dimension of a cyclic code. Namely, the set $\{g(t), tg(t), \ldots, t^{n-k-1}g(t)\}$ forms a basis for C, where $k = \deg g$. Hence, $\dim C = n - \deg g$.

We will assume throughout that $(n, p) = 1$. The roots of $g(t)$ in an algebraic closure $\overline{\mathbb{F}}_q$ of $\mathbb{F}_q$ are called *zeros* of the cyclic code C. Obviously the zeros of a cyclic code are the common roots of all the codewords in C. Since $g(t)$ divides $t^n - 1$ and the latter polynomial does not have multiple roots, the number of zeros of C is equal to the degree of $g(t)$. Hence the dimension of a cyclic code C of length n is $n - k$, where k is the number of zeros of C.

Assume that $t^n - 1 = g_1(t) \cdots g_s(t)$ is the factorization into irreducible polynomials over $\mathbb{F}_q$. Since the generator polynomial $g(t) \in \mathbb{F}_q[t]$ of a cyclic code C divides $t^n - 1$, it is a product of some combination of the factors $g_1, \ldots, g_s$. In particular, zeros of C are roots of the irreducible factors of $g(t)$. If α_i is a root of the irreducible polynomial g_i in some extension of $\mathbb{F}_q$, then the other roots of g_i are obtained by consecutive q^{th} powers of α_i, i.e. $\mathbb{F}_q$-conjugates. In order to describe a cyclic code over $\mathbb{F}_q$, it is enough to tell the irreducible factors of its generating polynomial. An irreducible polynomial is uniquely determined by its roots, or just one of its roots, as explained above. Hence, a q-ary cyclic code C is uniquely described by a subset of its zeros; namely a subset consisting of exactly one root of each irreducible factor of the generator polynomial $g(t)$. We call such a subset a *basic zero set* of C. Note that the choice of a basic zero set is in general not unique.

Example 3.2. Let C be a cyclic code of length n over $\mathbb{F}_q$, m be the order of $q \bmod n$, and α be a primitive n^{th} root of unity in $\mathbb{F}_{q^m}$. We call C a BCH *code of designed distance* d if the generator polynomial of C is the product of all distinct minimal polynomials over $\mathbb{F}_q$ of the $d - 1$ consecutive powers $\alpha^l, \alpha^{l+1}, \ldots, \alpha^{l+d-2}$. When $l = 1$, we call such a code a *narrow-sense* BCH code and if $n = q^m - 1$, i.e. α is a primitive element of $\mathbb{F}_{q^m}$, then the BCH code is called *primitive*.

An important example of a BCH code is obtained in the simplest case when $n = q - 1$. Then all zeros of the code lie in $\mathbb{F}_q$ or equivalently, all irreducible factors of the generating polynomial are linear. Namely, a *Reed-Solomon* (RS) code is a primitive BCH code of length $n = q - 1$. In this case, the generating polynomial has the form $\prod_{i=1}^{d-1}(t - \alpha^i)$, where α is a primitive element of $\mathbb{F}_q$. RS codes are commonly used in practice.

A well-known bound on the minimum distance of a cyclic code is the so-called *BCH bound* ([43, page 116]). It states that if the zero set of a cyclic code contains $\beta, \beta^2, \ldots, \beta^{d-1}$, where β is in some extension of $\mathbb{F}_q$ (e.g. a BCH code), then $d(C) \geq d$. This bound is valid for any cyclic code but it is in general weak. Still, it enables us to compute the minimum distance of RS codes and to prove that these codes are MDS as follows: Let C_{RS} be a RS code of length n with $d - 1$ consecutive zeros. Then, from the discussion above, we have

$$d(C_{RS}) \geq d \text{ and } k(C_{RS}) = n - (d - 1).$$

Hence, $d(C_{RS}) + k(C_{RS}) \geq n+1$ which implies, by the Singleton bound, that $d(C_{RS}) = d$.

In the following, we denote by Tr the trace mapping from $\mathbb{F}_{q^m}$ to $\mathbb{F}_q$. We also use the same symbol to denote the $\mathbb{F}_q$-linear map from $(\mathbb{F}_{q^m})^n$ to $\mathbb{F}_q^n$, which is defined by applying Tr componentwise.

Our aim is to state a minimum distance bound for cyclic codes which is much more effective than the BCH bound in general. One of the main ingredients in obtaining this result will be the following coding theoretic fact.

Theorem 3.3 (Delsarte). *For any code C over $\mathbb{F}_{q^m}$, we have*

$$(C|_{\mathbb{F}_q})^{\perp} = \mathrm{Tr}(C^{\perp}),$$

where $C|_{\mathbb{F}_q}$ denotes the restriction of C to $\mathbb{F}_q$, i.e. $C|_{\mathbb{F}_q} = C \cap \mathbb{F}_q^n$.

Proof. See [5] or [50, Theorem VIII.1.2]. □

Let $g(x) \in \mathbb{F}_{q^m}[x]$ be a polynomial, α be a primitive element of $\mathbb{F}_{q^m}$ and $n = q^m - 1$. We will use the following notation:

$$\Big(\mathrm{Tr}\big(g(x)\big)\Big)_{x \in \mathbb{F}_{q^m}^*} = \Big(\mathrm{Tr}\big(g(\alpha)\big), \mathrm{Tr}\big(g(\alpha^2)\big), \ldots, \mathrm{Tr}\big(g(\alpha^n)\big)\Big). \quad (3.2)$$

For polynomials $f_1(x), \ldots, f_t(x) \in \mathbb{F}_{q^m}[x]$, the $\mathbb{F}_{q^m}$-linear space spanned by the set $\{f_1, \ldots, f_t\}$ will be denoted by $\langle f_1, \ldots, f_t \rangle_{\mathbb{F}_{q^m}}$.

The following is the key result which relates cyclic codes to Artin-Schreier type equations.

Theorem 3.4 (Wolfmann [61]). *Let $m > 1$ and C be a q-ary cyclic code of length $n = q^m - 1$. Let α be a primitive element of $\mathbb{F}_{q^m}$ and $\{\alpha^{i_1}, \alpha^{i_2}, \ldots, \alpha^{i_s}\}$ be a basic zero set of the dual code $C^{\perp}$. Then*

$$C = \left\{ \Big(\mathrm{Tr}\big(f(x)\big)\Big)_{x \in \mathbb{F}_{q^m}^*} ; f(x) \in \langle x^{i_1}, \ldots, x^{i_s} \rangle_{\mathbb{F}_{q^m}} \right\}. \quad (3.3)$$

Proof. Viewing $C^{\perp}$ as an ideal in $\mathbb{F}_q[t]/(t^n - 1)$, note that

$$C^{\perp} = \big(f_{\alpha^{i_1}}(t) f_{\alpha^{i_2}}(t) \cdots f_{\alpha^{i_s}}(t)\big),$$

where $f_{\alpha^{i_j}}(t)$ is the minimal polynomial of α^{i_j} over $\mathbb{F}_q$, for all j. Let $\tilde{C}$ be the code over $\mathbb{F}_{q^m}$ of the same length whose dual has the zero set $\{\alpha^{i_1}, \alpha^{i_2}, \ldots, \alpha^{i_s}\}$, i.e.

$$\tilde{C}^{\perp} = \big((t - \alpha^{i_1})(t - \alpha^{i_2}) \cdots (t - \alpha^{i_s})\big) \subset \mathbb{F}_{q^m}[t]/(t^n - 1).$$

Then $C^{\perp}$ is the restriction of $\tilde{C}^{\perp}$ to $\mathbb{F}_q$ and hence we have $C = \mathrm{Tr}(\tilde{C})$ by Delsarte's theorem.

Observe that for any codeword $b(t) = \sum_{i=0}^{n-1} b_i t^i$ in $\tilde{C}^\perp$, we have $b(\alpha^{i_j}) = 0$, for all $j = 1, 2, \ldots, s$. These equalities can also be written using the usual inner product in n-space as follows:

$$(b_0, b_1, \ldots, b_{n-1}) \cdot (1, (\alpha^{i_1})^1, \ldots, (\alpha^{i_1})^{n-1}) = 0,$$
$$(b_0, b_1, \ldots, b_{n-1}) \cdot (1, (\alpha^{i_2})^1, \ldots, (\alpha^{i_2})^{n-1}) = 0,$$
$$\vdots$$
$$(b_0, b_1, \ldots, b_{n-1}) \cdot (1, (\alpha^{i_s})^1, \ldots, (\alpha^{i_s})^{n-1}) = 0.$$

Remembering the vector representation of cyclic codes, the above equalities mean that the following vectors are codewords in $\tilde{C}$:

$$\left.\begin{array}{l} v_1 = (1, (\alpha^1)^{i_1}, \ldots, (\alpha^{n-1})^{i_1}) = (x^{i_1})_{x \in \mathbb{F}_{q^m}^*} \\ v_2 = (1, (\alpha^1)^{i_2}, \ldots, (\alpha^{n-1})^{i_2}) = (x^{i_2})_{x \in \mathbb{F}_{q^m}^*} \\ \vdots \\ v_s = (1, (\alpha^1)^{i_s}, \ldots, (\alpha^{n-1})^{i_s}) = (x^{i_s})_{x \in \mathbb{F}_{q^m}^*} \end{array}\right\} \in \tilde{C}.$$

The generator polynomial of $\tilde{C}^\perp$ reveals that the $\mathbb{F}_{q^m}$-dimension of $\tilde{C}$ is s. It is not difficult to show that $\{v_1, \ldots, v_s\}$ is a linearly independent set over $\mathbb{F}_{q^m}$. Hence, it forms an $\mathbb{F}_{q^m}$-basis for $\tilde{C}$. This implies that any codeword in $C = \mathrm{Tr}(\tilde{C})$ can be written as $\mathrm{Tr}(\lambda_1 v_1 + \cdots + \lambda_s v_s)$ for some λ_j in $\mathbb{F}_{q^m}$. $\square$

Corollary 3.5 (Wolfmann [61]). *Let $m > 1$ and C be a q-ary cyclic code of length $n = q^m - 1$. Let α be a primitive element of $\mathbb{F}_{q^m}$ and $\{\alpha^{i_1}, \alpha^{i_2}, \ldots, \alpha^{i_s}\}$ be a basic zero set of the dual code $C^\perp$, where $0 < i_1 < \cdots < i_s$. If*

$$\gcd(i_j, p) = 1, \quad for \ all \ j = 1, \ldots, s \tag{3.4}$$

then the nonzero weights w of C satisfy

$$\left| w - (q^m - q^{m-1}) \right| \leq (q-1)(i_s - 1)q^{\frac{m}{2}-1}. \tag{3.5}$$

Proof. Let c be a nonzero codeword in C with a trace representation as in Theorem 3.4 with some $(\lambda_1, \ldots, \lambda_s) \in \mathbb{F}_{q^m}^s$. By Hilbert's Theorem 90 (cf. (2.3)), the weight of c is

$$w(c) = q^m - \frac{N}{q}, \tag{3.6}$$

where N is the number of solutions of the following equation in $\mathbb{F}_{q^m} \times \mathbb{F}_{q^m}$:

$$y^q - y = \lambda_1 x^{i_1} + \lambda_2 x^{i_2} + \cdots + \lambda_s x^{i_s} \tag{3.7}$$

By assumption (3.4) and Example 2.4, this equation is irreducible over the rational function field $\mathbb{F}_{q^m}(x)$. Hence it defines a degree q extension of $\mathbb{F}_{q^m}(x)$

with genus $g = (q-1)(\deg f - 1)/2$ (cf. Equation (2.2)), where f denotes the polynomial on the right hand side of Equation (3.7). Using Hasse-Weil bound for the largest possible genus, i.e. $\deg f = i_s$, the result follows.   □

Remark 3.6. Assumption (3.4), which is required to guarantee irreducibility of Equation (3.7) over $\mathbb{F}_{q^m}(x)$, need not hold for every q-ary cyclic code. However if $q = p$, then there is a choice of a basic zero set for $C^\perp$ which necessarily satisfies this condition. Note that if a and b are in the same p-cyclotomic coset modulo $n = p^m - 1$, i.e. $a \equiv bp^u \pmod{n}$ for some u, then α^a and α^b are roots of the same irreducible polynomial in $\mathbb{F}_p[t]$. If each i_j is chosen as the minimal element of its cyclotomic coset, then it is necessarily relatively prime to p.

Note that in any case one should choose a basic zero set consisting of smallest possible numbers, satisfying (3.4), in order to lower the genus term in the bound and hence obtain the strongest result.

Remark 3.7. Let n be a proper divisor of $q^m - 1$, ω be a primitive n^{th} root of unity, and D be a cyclic code of length n whose dual $D^\perp$ has $\{\omega^{i_1}, \ldots, \omega^{i_s}\}$ as a basic zero set. Then the above proof also works for D and every codeword can be represented as $(\mathrm{Tr}(f(\omega^0)), \mathrm{Tr}(f(\omega^1)), \ldots, \mathrm{Tr}(f(\omega^{n-1})))$ with some $f(x) = \lambda_1 x^{i_1} + \cdots + \lambda_s x^{i_s}$ in $\mathbb{F}_{q^m}[x]$. Here, we denote again the trace map $\mathrm{Tr}_{\mathbb{F}_{q^m}/\mathbb{F}_q}$ by Tr for simplicity. Now, let α be a primitive $(q^m - 1)^{th}$ root of unity and C be a cyclic code of length $q^m - 1$ with the basic zero set $\{\alpha^{ri_1}, \ldots, \alpha^{ri_s}\}$ for its dual $C^\perp$, where $r = (q^m - 1)/n$. In this case a codeword in C has the form $(\mathrm{Tr}(\bar{f}(\alpha^0)), \mathrm{Tr}(\bar{f}(\alpha^1)), \ldots, \mathrm{Tr}(\bar{f}(\alpha^{q^m-2})))$ with some $\bar{f}(x) = \lambda_1 x^{ri_1} + \cdots + \lambda_s x^{ri_s}$ in $\mathbb{F}_{q^m}[x]$. Then it is easy to see that $d(D) = \frac{n}{q^m-1} d(C)$. This shows that one can, without loss of generality, study only the weights of cyclic codes of "full length" $q^m - 1$.

Let $i_1, \ldots, i_s$ be nonnegative integers, $r_1, \ldots, r_s$ be positive integers which are not divisible by p and $\lambda_1, \ldots, \lambda_s \in \mathbb{F}_{q^m}$. Consider the equation

$$y^q - y = \lambda_1 x^{r_1 p^{i_1}} + \cdots + \lambda_s x^{r_s p^{i_s}}. \tag{3.8}$$

By Theorem 2.3 (iii), this equation is reducible over $\mathbb{F}_{q^m}(x)$ if and only if there exists $a \in \mathbb{F}_q^*$ such that

$$y^p - y = a\lambda_1 x^{r_1 p^{i_1}} + \cdots + a\lambda_s x^{r_s p^{i_s}} \tag{3.9}$$

is reducible over $\mathbb{F}_{q^m}(x)$. Note that for any $1 \le j \le s$, the change of variable $y \leftrightarrow (y - (a\lambda_j)^{p^{-1}} x^{r_j p^{i_j-1}})$ in (3.9) does not change the function field defined by this equation. However, if it is applied for all $j = 1, \ldots, s$ as many times as necessary, it reduces (3.9) to

$$y^p - y = (a\lambda_1)^{p^{-i_1}} x^{r_1} + \cdots + (a\lambda_s)^{p^{-i_s}} x^{r_s}. \tag{3.10}$$

If there exists r_j which is different than all the other $r_{j'}$'s, then for any $a \in \mathbb{F}_q^*$ the term x^{r_j} in (3.10) will be isolated and it will have a nonzero coefficient. Hence, by (2.1) and Theorem 2.3 (iii,iv), we have the following:

Proposition 3.8. *Consider*

$$y^q - y = \lambda_1 x^{r_1 p^{i_1}} + \cdots + \lambda_s x^{r_s p^{i_s}},$$

where $\lambda_j \in \mathbb{F}_{q^m}$ and $p \nmid r_j$ for all $j = 1, 2, \ldots, s$. If there exists r_j such that $r_j \neq r_{j'}$ for any $j' \neq j$, then the equation defines an Artin-Schreier type extension of $\mathbb{F}_{q^m}(x)$. In this case the genus of the extension satisfies $g \leq (q-1)(r-1)/2$, where $r = \max\{r_1, \ldots, r_s\}$. If $r_1, \ldots, r_s$ are pairwise distinct, then $g = (q-1)(r-1)/2$.

Note that this proposition extends the conclusions in Example 2.4 to more general Artin-Schreier type equations. Hence, we can relax the condition (3.4) in Corollary 3.5 and obtain an immediate extension of Wolfmann's result to a class of cyclic codes for which the dual's basic zero set is of the form

$$\{\alpha^{r_1 p^{i_1}}, \ldots, \alpha^{r_s p^{i_s}}\}, \tag{3.11}$$

where $r_1, \ldots, r_s$ are pairwise distinct.

In [27], the following Hasse-Weil type bound is obtained for the number of solutions in $\mathbb{F}_{q^m} \times \mathbb{F}_{q^m}$ of a reducible Artin-Schreier type equation.

Theorem 3.9. *Let $q = p^e$ and $r_1, \ldots, r_k$ be pairwise distinct positive integers which are not divisible by p. Let $0 \leq i_{t,1} < i_{t,2}$ and $j_t = i_{t,2} - i_{t,1}$ for $t = 1, \ldots, k$. Consider a reducible equation*

$$y^q - y = \left(\lambda_1 x^{r_1 p^{i_{1,1}}} + \beta_1 x^{r_1 p^{i_{1,2}}}\right) + \cdots + \left(\lambda_k x^{r_k p^{i_{k,1}}} + \beta_k x^{r_k p^{i_{k,2}}}\right), \tag{3.12}$$

where $\lambda_1, \ldots, \lambda_k, \beta_1, \ldots, \beta_k \in \mathbb{F}_{q^m}$. Assume that $r = \max\{r_1, \ldots, r_k\}$, $l = \gcd(e, j_1, \ldots, j_k)$ and N denotes the number of solutions of (3.12) in $\mathbb{F}_{q^m} \times \mathbb{F}_{q^m}$. Then we have

$$|N - p^{em+l}| \leq (p^e - p^l)(r-1)\sqrt{p^{em}}.$$

The consequence of this theorem for cyclic codes is clear: it allows us to extend Corollary 3.5 further. Namely, one can allow pairs of equal r_i's to appear in the dual's basic zero set in (3.11). Details and the bound for the weights of cyclic codes can be found in [27]. Naturally, finding results similar to Theorem 3.9 for the number of solutions of more general reducible equations would extend Wolfmann's approach and his bound to even wider classes of cyclic codes.

Example 3.10. Let $q = 2^m$ for some $m > 2$, α be a primitive element in $\mathbb{F}_q$, and consider the binary BCH code B_m of length $n = q - 1$ with the generating polynomial $g(t) = f_\alpha(t)f_{\alpha^3}(t) \in \mathbb{F}_q[t]$, i.e. $\{\alpha, \alpha^3\}$ is a basic zero set. The code B_m is also referred to as *double-error-correcting* binary BCH code and denoted as BCH(2). For any $m > 2$, the zero set of B_m contains $\alpha, \alpha^2, \alpha^3, \alpha^4$. Hence, $d(B_m) \geq 5$ by the BCH bound. In fact, this is the exact minimum distance and we can say more about the weights of B_m.

Note that the dual code can be represented as

$$B_m^\perp = \left\{ \left(\mathrm{Tr}(f(x))\right)_{x \in \mathbb{F}_q^*};\ f(x) \in \langle x, x^3 \rangle_{\mathbb{F}_q} \right\}.$$

Hence, weights in $B_m^\perp$ are related to the family $\mathcal{F} = \{y^2 - y = \lambda x + \mu x^3;\ \lambda, \mu \in \mathbb{F}_q\}$. If $\mu = 0$, then the resulting function field is rational and the defining equation has exactly q solutions in $\mathbb{F}_q \times \mathbb{F}_q$. When $\mu \neq 0$, we have a genus 1 (elliptic) function field (cf. Example 2.4). In this case, the possible number of $\mathbb{F}_q$-rational points and their frequencies in $\mathcal{F}$ are known (see [45]). Hence, we have the list of all the weights and their frequencies for the code $B_m^\perp$, i.e. the so-called *weight enumerator* of the code. Then the weight enumerator of the BCH code B_m can be obtained using the *MacWilliams identity* in coding theory ([36, Page 41]), which gives a relation between the weight enumerators of dual codes (see [46] for details).

We have demonstrated two research directions related to the problem in hand. Wolfmann's work, together with Theorem 3.9, covers a large class of cyclic codes whereas Example 3.10 studies a specific family of codes. Although one naturally obtains stronger results in special cases, e.g. complete weight enumerator as opposed to a minimum distance bound, this approach requires a more in-depth study of the related family of curves. The problem becomes particularly difficult when the genus is large. We will refer to some works in the literature that utilize the method described here, i.e. algebraic curves via trace representation.

- Binary Melas codes and their duals are studied in [47]. The ternary case for these codes is considered in [14, 21]. In both cases, complete weight enumerators are obtained. Related families of curves for the dual Melas codes are

$$\text{binary}: \quad \mathcal{F}_b = \{y^2 - y = ax + b/x;\ a, b \in \mathbb{F}_{2^m}\}$$
$$\text{ternary}: \quad \mathcal{F}_t = \{y^3 - y = ax + b/x;\ a, b \in \mathbb{F}_{3^m}\}$$

In other words, a basic zero set for the Melas code M_m of length $2^m - 1$ (resp. $3^m - 1$) is $\{\alpha, \alpha^{-1}\}$, where α is a primitive element of $\mathbb{F}_{2^m}$ (resp. $\mathbb{F}_{3^m}$).

- In [37], the intersection of the binary Melas code M_m and the binary double-error-correcting BCH code B_m (cf. Example 3.10) is studied for $m > 2$. In other words, if α is a primitive element of $\mathbb{F}_{2^m}$, then the generating polynomial of the code C is $f_\alpha(x) f_{\alpha^{-1}}(x) f_{\alpha^3}(x)$ and the related family of curves for $C^\perp$ is $\mathcal{F} = \{y^2 - y = ax + b/x + cx^3; \ a, b, c \in \mathbb{F}_{2^m}\}$. The set of possible weights for $C^\perp$ (not the weight enumerator) is obtained exactly for the case m even. For m odd, a less complete but similar result is found.

- Let C be the binary, narrow-sense, primitive BCH code of length $2^m - 1$ with designed distance $d = 2t+1$, i.e. the generating polynomial of C is the least common multiple of $f_\alpha(x), f_{\alpha^2}(x), \ldots, f_{\alpha^{d-1}}(x)$, where α is a primitive element of $\mathbb{F}_{2^m}$. Note that for any element α^j, there is an odd number $\tilde{j}$ such that α^j and $\alpha^{\tilde{j}}$ share the same minimal polynomial over $\mathbb{F}_2$, i.e. they are $\mathbb{F}_2$-conjugate. Hence, the generating polynomial of C is the least common multiple of the minimal polynomials $f_\alpha(x), f_{\alpha^3}(x), \ldots, f_{\alpha^{2t-1}}(x)$ over $\mathbb{F}_2$. This also shows that choosing $d = 2t + 1$ rather than $d = 2t$ does not cause a lack of generality since these codes coincide. Since a basic zero set consisting only of odd powers of α can be chosen, Corollary 3.5 can be used for $C^\perp$. In fact, it yields $|w - 2^{m-1}| \le (t-1)\sqrt{2^m}$ for the nonzero weights w of the dual BCH code $C^\perp$ ($q = 2$ and $i_s = 2t - 1$ in (3.5)). Note that this bound is nontrivial for $t - 1 < 2^{m/2-1}$. In [59], the author improves the Hasse-Weil bound on the related family of curves for $C^\perp$ using a variant of the method in [53]. This way, he obtains an effective bound for the weights of $C^\perp$ in an extended range of t values.

Remark 3.11. Let $n_1, \ldots, n_l$ be positive integers which are relatively prime to p. By definition, an $l - D$ *cyclic code* over $\mathbb{F}_q$ of *volume* $n_1 \times \cdots \times n_l$ is an ideal of the ring $R = \mathbb{F}_q[x_1, \ldots, x_l]/(x_1^{n_1} - 1, \ldots, x_l^{n_l} - 1)$. Analogy with cyclic codes is clear but it can be made more transparent in the case of $l = 2$. Denote the $\mathbb{F}_q$-linear space of dimension $n_1 n_2$ by $\mathbb{F}_q^{n_1 \times n_2}$ and represent the vectors in this space as $n_1 \times n_2$ arrays. As in the case of cyclic codes, we have the following isomorphism:

$$\mathbb{F}_q^{n_1 \times n_2} \quad \longleftrightarrow \quad \mathbb{F}_q[x, y]/(x^{n_1} - 1, y^{n_2} - 1)$$

$$(a_{i,j}) \quad \longleftrightarrow \quad \sum_{i=0}^{n_1-1} \sum_{j=0}^{n_2-1} a_{i,j} x^i y^j$$

Note that under this identification a $2 - D$ cyclic code is an $\mathbb{F}_q$-subspace of $\mathbb{F}_q^{n_1 \times n_2}$ which is closed under row and column shifts of codewords (arrays).

As in the case of cyclic codes, one can start with a "basic zero set" of the dual $l - D$ cyclic code $C^\perp$ to obtain a trace representation for the code C. The difference with cyclic codes is that the trace expressions involve polynomials in l indeterminates. Hence, the weight of a codeword is determined by the number of rational points of a higher dimensional geometric object; namely an Artin-Schreier hypersurface of the form $y^q - y = f(x_1, \ldots, x_l)$. In this case, one can again use the Hasse-Weil bound for curves or a bound on character sums in several variables due to Deligne (see [4, Proposition 3.8]). The results and further information can be found in [24, 26].

For an interesting application of the above-mentioned bound of Deligne to minimum distance analysis of certain cyclic codes, we refer to [38].

4.   Trace Codes

In this section we will introduce the general class of codes that can be represented by means of the trace map. These codes are called *trace codes*. We saw in the previous section that any cyclic code has a trace representation (Theorem 3.4). Hence, cyclic codes will be a special case of the type of codes to be investigated in this section (cf. Remark 4.3). Our main interest will be the weights, particularly the so-called *higher weights* (*generalized Hamming weights*), of a class of trace codes obtained from function fields.

As in the previous section, Tr will denote the trace map from $\mathbb{F}_{q^m}$ to $\mathbb{F}_q$ as well as the induced $\mathbb{F}_q$-linear map from $(\mathbb{F}_{q^m})^n$ to $(\mathbb{F}_q)^n$. When $q = p$, we will denote both maps by tr.

Definition 4.1. Let C be a code of length n over $\mathbb{F}_{q^m}$. The trace code $\mathrm{Tr}(C)$ of C is defined by $\mathrm{Tr}(C) = \{\mathrm{Tr}(c) : c \in C\}$.

Note that $\mathrm{Tr}(C)$ is an $\mathbb{F}_q$-linear code and it is dual to the restriction code (or the subfield subcode) $(C^\perp)|_{\mathbb{F}_q}$ of $C^\perp$, by Delsarte's Theorem. For an $\mathbb{F}_{q^m}$-linear code C of length n, the following estimates on the dimensions of its subfield subcode and its trace code follow from the definitions and Delsarte's theorem (see [50, Lemma VIII.1.3]).

$$\dim C - (m-1)(n - \dim C) \leq \dim C|_{\mathbb{F}_q} \leq \dim C \leq \dim \mathrm{Tr}(C) \leq m \dim C.$$
$$(4.1)$$

We note that the dimensions above represent dimensions over different fields, depending on the alphabets of the codes considered, e.g. $\dim C$ is considered over $\mathbb{F}_{q^m}$ whereas $\dim \mathrm{Tr}(C)$ is considered over $\mathbb{F}_q$. We refer the interested reader to [49, 56–58] for improvements on these general dimension bounds for some special classes of codes.

Example 4.2. We are interested in trace codes obtained from algebraic function fields. Namely, let F be a function field with full constant field $\mathbb{F}_{q^m}$ and $V \subset F$ be a finite dimensional $\mathbb{F}_{q^m}$-subspace. For distinct rational places $P_1, \ldots, P_n$ of F, let $D = P_1 + \cdots + P_n$ and assume that $v_{P_i}(f) \geq 0$ for all $f \in V$. Then we define

$$C_D(V) = \{(f(P_1), \ldots, f(P_n)) : f \in V\} \subset (\mathbb{F}_{q^m})^n \qquad (4.2)$$

and

$$\text{Tr}_D(V) = \text{Tr}(C_D(V)) = \{(\text{Tr}(f(P_1)), \ldots, \text{Tr}(f(P_n))) : f \in V\} \subset (\mathbb{F}_q)^n \tag{4.3}$$

Remark 4.3. Note that if $F = \mathbb{F}_{q^m}(x)$ is the rational function field, the set $\{P_1, \ldots, P_n\}$ consists of the set of all rational places of F except the the zero and the pole of x, and V is the $\mathbb{F}_{q^m}$-linear space spanned by the set of functions $x^{i_1}, \ldots, x^{i_s}$ in F, then we obtain the cyclic code in Equation (3.3).

We also note that the codes in (4.2) are generalizations of *algebraic geometry (AG) codes*. Namely, if $V = L(G)$ is the Riemann-Roch space of a divisor G whose support does not contain any of the P_i's, then the resulting code $C_D(L(G))$ is an AG code. We refer to [50, Chapters 2,7] for background on AG codes.

It is clear that as in the case of cyclic codes, the weights of codewords in $\text{Tr}_D(V)$ are related to Artin-Schreier type equations via Hilbert's Theorem 90 (see [50, Chapter 8]). Our aim is to extend this relation to the so-called higher weights of the code. For this we need some definitions.

Definition 4.4. Let $\mathbb{F}$ be a finite field. For a subset S of $\mathbb{F}^n$, we define the *support* and *weight* of S by

$$\text{Supp}(S) = \{i : \exists (s_1, \ldots, s_n) \in S \text{ with } s_i \neq 0\}, \quad w(S) = |\text{Supp}(S)|.$$

Definition 4.5. Let C be an $[n, k]$ code over the finite field $\mathbb{F}$. For $1 \leq r \leq k$, we define the rth *generalized Hamming weight* of C by

$$d_r(C) = \min\{w(S) : S \text{ is an } r\text{-dimensional subcode of } C\}.$$

We call the set $\{d_1(C), \ldots, d_k(C)\}$ the *(higher) weight hierarchy* of C. Note that $d_1(C)$ is nothing but the minimum distance of C. It is also clear that the weight hierarchy is a nondecreasing positive sequence. In fact more is true.

Proposition 4.6. *For an $[n, k]$ code C over $\mathbb{F}$, we have*
(i) $0 < d_1(C) < d_2(C) < \cdots < d_k(C) \leq n$.
(ii) (Generalized Singleton Bound) $r \leq d_r(C) \leq r + (n - k)$, *for all* r.

Proof. (i) We want to show that the strict inequality $d_{r-1}(C) < d_r(C)$ holds for any $2 \leq r \leq k$. Let $S \subseteq C$ be an r-dimensional subcode with $w(S) = d_r(C)$. Let i be an element in $\text{Supp}(S)$ and consider the projection $p_i : S \to \mathbb{F}$ sending $(s_1, \ldots, s_n)$ to s_i. This is an $\mathbb{F}$-linear map which is onto, since $i \in \text{Supp}(S)$. Therefore the kernel $\tilde{S}$ is an $(r-1)$-dimensional subcode of C and $i \notin \text{Supp}(\tilde{S})$. This implies that $d_{r-1}(C) \leq w(\tilde{S}) < w(S) = d_r(C)$.

(ii) The monotonicity result obtained in part (i) immediately yields $r \leq d_r(C)$. Again by part (i) we have $(k-r) \leq (d_k(C) - d_r(C))$. Since $d_k(C) \leq n$

the second inequality, which yields the Singleton bound for the case $r = 1$, follows.   □

Generalized Hamming weights were introduced by Wei in [60]. Besides proving the results in Proposition 4.6, he computed weight hierarchies of certain well-known codes and proved a "duality theorem" between the weight hierarchies of C and $C^{\perp}$:

$$\{d_r(C) : 1 \le r \le k\} = \{1, 2, \ldots, n\} \setminus \{n + 1 - d_r(C^{\perp}) : 1 \le r \le n - k\}.$$

Later, further relations between the generalized Hamming weights of dual codes were obtained. For $1 \le i \le n$ and $1 \le r \le k$, define

$$A_i^r(C) = |\{S : S \text{ is a linear subspace of } C \text{ with } \dim S = r \text{ and } w(S) = i\}|.$$

The set $\{A_i^r(C) : 1 \le i \le n, \ 1 \le r \le k\}$ is called the *support (effective length) distribution* of C. This set is also referred to as the *higher weight spectrum* of C. Note that this is a generalization of the weight enumerator (cf. Example 3.10). The MacWilliams identity states that the weight enumerators of dual codes completely determine each other. The same is also true for support weight distributions of dual codes. MacWilliams-type identitites for generalized Hamming weights were found by Kløve in [32] and Simonis in [48].

After Wei's article, the subject developed in two directions: investigation of general properties of higher weights and generalized weight studies of particular families of codes. We tried to mention some of the general results above. For more information, we refer to [30, 55]. At the end of this section we will mention some of the works that use function field theoretic approach in the higher weights study of specific classes of codes.

The main aim in this section is to find a relation between generalized weights of trace codes obtained from function fields and the number of rational places of certain algebraic function fields. This relation was established in [51] and [16]. We will give a proof of this result for p-ary trace codes (see the comments on the q-ary case in Remark 4.11). Therefore, we replace q by p in Example 4.2 and consider a function field F whose full constant field is $\mathbb{F}_{p^m}$. The rest of the ingredients of Example 4.2, i.e. V and D, are also considered accordingly. Hence, the trace code obtained in (4.3) is a p-ary code and, following our notational convention, it is denoted as follows:

$$C = \mathrm{tr}_D(V) = \mathrm{tr}(C_D(V)) = \{(\mathrm{tr}(f(P_1)), \ldots, \mathrm{tr}(f(P_n))) : f \in V\} \subset (\mathbb{F}_p)^n \tag{4.4}$$

Definition 4.7. We call a subspace U of V an *r-dimensional D-regular subspace* of V if $\dim_{\mathbb{F}_p} U = r$ and $\mathrm{tr}_D(f)$ is not equal to $(0, \ldots, 0)$ or $(1, \ldots, 1)$, for any nonzero $f \in U$.

Proposition 4.8. *Let $f \in F$ be an element such that $\mathrm{tr}_D(f) \neq (0, \ldots, 0)$. Then $\psi_f(T) = T^p - T - f$ is irreducible over F.*

Proof. Let us start by observing a well-known fact which is true without the assumptions we made on f. The polynomial ψ_f is either irreducible over F or it splits into linear factors in $F[T]$. Note that if α is a root of ψ_f, then the roots of ψ_f are $\alpha + c$, where $c \in \mathbb{F}_p$. Hence, if one root lies in F, which contains $\mathbb{F}_p$, all the roots lie in F. Assume that ψ_f does not split into linear factors over F and it is not irreducible over F. Write $\psi_f(T) = \beta(T)\mu(T)$ for $\beta, \mu \in F[T]$ with $0 < d = \deg \beta < p$. We know that ψ_f splits into linear factors, with roots $\{\alpha + c : c \in \mathbb{F}_p\}$, in some extension of F. Therefore the coefficient of T^{d-1} in β is of the form $d\alpha + k$ for some positive integer k. Since β has coefficients in F and $d \neq 0$, we conclude $\alpha \in F$, which is a contradiction. Hence, ψ_f must be irreducible over F.

Now, assume that ψ_f is reducible over F. Then, by the above fact, there exists a root $z \in F$, i.e. $z^p - z = f$. Since $v_{P_i}(f) \geq 0$ for all $1 \leq i \leq n$, we also have $v_{P_i}(z) \geq 0$ for all $1 \leq i \leq n$ (using the triangle inequality). Hence $z(P_i)$ makes sense and we have

$$
\begin{aligned}
\mathrm{tr}_D(f) &= \big(\mathrm{tr}(f(P_1)), \ldots, \mathrm{tr}(f(P_n))\big) \\
&= \big(\mathrm{tr}(z(P_1)^p - z(P_1)), \ldots, \mathrm{tr}(z(P_n)^p - z(P_n)))\big) \\
&= (0, \ldots, 0).
\end{aligned}
$$

This contradicts the assumption. $\qquad\square$

For an r-dimensional D-regular subspace U of V, let E_U be the splitting field of all the irreducible polynomials ψ_f (as f runs through U) over F. Note that E_U/F is a Galois extension since ψ_f is separable for all $f \in U$. In fact, E_U/F is an elementary abelian extension of degree p^r (see [11, Section 1] or [51, Proposition 1]). If $\{f_1, \ldots, f_r\}$ is an $\mathbb{F}_p$-basis of U and $\psi_{f_i}(y_i) = 0$ for $1 \leq i \leq r$, then it can be seen that $E_U = F(y_1, \ldots, y_r)$. Furthermore, the full constant field of the extension E_U is also $\mathbb{F}_{p^m}$. We note that the validity of these assertions rely on the regularity assumption on U.

Proposition 4.9. *Let $\mathbb{P}_F^{(1)}$ denote the set of rational places of $F/\mathbb{F}_{p^m}$ and $R = \mathbb{P}_F^{(1)} \setminus \{P_1, \ldots, P_n\}$. For an r-dimensional D-regular subspace $U \subset V$, let E_U be the extension of F as defined above and define the set*

$$
\mathrm{Supp}(U) = \{P_i : 1 \leq i \leq n \text{ and } \mathrm{tr}(f(P_i)) \neq 0 \text{ for some } f \in U\},
$$

whose cardinality is denoted by $w(U)$. Let $\overline{R}_U = \{Q \in \mathbb{P}_{E_U}^{(1)} : Q \cap F \in R\}$. Then the number of rational places $N(E_U)$ of E_U satisfies

$$
N(E_U) = p^r(n - w(U)) + |\overline{R}_U|. \tag{4.5}
$$

Proof. A rational place of E_U lies over a rational place of F. Since the set $\overline{R}_U$ collects those rational places lying over R, we need to check the extensions of the places $P_1, \ldots, P_n$. If $P_i \in \text{Supp}(U)$, then there exists $f \in U$ with $\text{tr}(f(P_i)) \neq 0$. Consider the irreducible polynomial $\psi_f(T) = T^p - T - f$ over F and the Artin-Schreier intermediate field $F \subset F(y_f) \subset E_U$ that it defines for some $y_f \in E_U$ with $\psi_f(y_f) = 0$. Since $\text{tr}(f(P_i)) \neq 0$, the reduction of ψ_f mod P_i is irreducible over $\mathbb{F}_{p^m}$ which implies, by Kummer's Theorem (see [50, Theorem III.3.7]), that P_i has a unique extension of degree p in $F(y_f)$. Therefore, for those P_i that lie in $\text{Supp}(U)$, there is no rational extension in E_U. If $P_i \notin \text{Supp}(U)$, then $\text{tr}(f(P_i)) = 0$ for all $f \in U$. Hence, again by Kummer's Theorem, P_i has p rational extensions in each intermediate field $F(y_f)$ (as f runs through U). Since E_U is the splitting field of $F(y_f)$'s, using [50, Corollary III.8.4], P_i has p^r rational extensions in E_U. Hence, Equation (4.5) follows.   □

Let S be an r-dimensional subcode of the code C in (4.4) and $(1, \ldots, 1) \notin S$. If $\{s_1, \ldots, s_r\}$ is an $\mathbb{F}_p$-basis of S, let $f_i \in V$ such that $\text{tr}_D(f_i) = s_i$, for $i = 1, \ldots, r$. Consider $U_S = \langle f_1, \ldots, f_r \rangle_{\mathbb{F}_p} \subset V$. It is easy to see that U_S is an r-dimensional D-regular subspace of V such that $\text{tr}_D(U_S) = S$. Also note that, with the notation of Proposition 4.9, we have $w(S) = w(U_S)$. Hence, we get the following:

Theorem 4.10. *With the notation as above assume that* $(1, \ldots, 1) \notin C = \text{tr}_D(V)$. *Then for the* r^{th} *generalized Hamming weight of* C *we have*

$$d_r(C) = n - \frac{1}{p^r} \max \left(N(E_{U_S}) - |\overline{R}_{U_S}| \right), \qquad (4.6)$$

where the maximum is taken over all r-dimensional subcodes S of C.

Remark 4.11. A natural thing to do is to obtain estimates on d_r by bounding the term $\max \left(N(E_{U_S}) - |\overline{R}_{U_S}| \right)$. If one attempts to use the Hasse-Weil bound for this purpose, then the genus computation (or estimation) of E_{U_S} is required. If F is the rational function field $\mathbb{F}_{p^m}(x)$, then this can be done by Theorem 2.3(iv). Otherwise, one needs to use a more general genus formula which is good for elementary abelian extensions of arbitrary function fields (see [11, Theorem 2.1]).

We refer to [25] for the study of q-ary trace codes ($q \neq p$). This case requires a new notion, called strongly linearly independent subspace, that plays the role of a regular subspace in the p-ary case.

Finally, both in the weight analysis in this section (Theorem 4.10) and in Section 3 (Corollary 3.5 and Theorem 3.9), one could use an improvement of the Hasse-Weil bound due to Serre ([50, Theorem V.3.1]) to estimate the number of rational places of function fields. Serre's improvement replaces the part $2g\sqrt{q}$ in the Hasse-Weil bound by $\lfloor 2\sqrt{q} \rfloor g$. In [39], an improvement of Serre's bound for elementary abelian extensions of the rational function field is employed to improve the generalized weight estimate on the dual of a p-ary BCH code, which is obtained in [51] (see Example 4.12). In [25], the technique used by Moreno *et al* in [39], called "p-divisibility", is extended to "q-divisibility", and it is combined with other methods to obtain further improvements of Serre's bound for elementary abelian extensions. This is then applied to a large class of q-ary trace codes. We do not include these estimates on the number of rational places since they are rather technical.

Example 4.12. Let $C = \text{BCH}(t)$ be the binary, narrow-sense, primitive BCH code of length $n = 2^m - 1$ with designed distance $d = 2t + 1$. We saw in Section 3 that the generating polynomial of C is the least common multiple of the minimal polynomials $f_\alpha(x), f_{\alpha^3}(x), \ldots, f_{\alpha^{2t-1}}(x)$ over $\mathbb{F}_2$, where α is a primitive element of $\mathbb{F}_{2^m}$. We want to analyze generalized weights of the dual code $C^\perp$. For an estimate on d_r of cyclic codes, in general, we refer to [51, Theorem 10].

Since $(1, \ldots, 1) \in C$ and n is odd, $(1, \ldots, 1) \notin C^\perp$. Note that, with the notation of Example 4.2 and Remark 4.3, $C^\perp = \text{tr}_D(V)$, where $D = P_1 + \cdots + P_n$ is the sum of all rational places of $F = \mathbb{F}_{2^m}(x)$ except the zero P_0 and the pole P_∞ of x, and $V = \langle x, x^3, \ldots, x^{2t-1} \rangle_{\mathbb{F}_{2^m}} \subset F$. For an r-dimensional subcode S of $C^\perp$, consider the extension E_{U_S}/F. Since an element $f \in U_S$ is of the form $f = c_1 x + c_2 x^3 + \cdots + c_t x^{2t-1}$ for some $c_i \in \mathbb{F}_{2^m}$, the genus of the intermediate field $E_f = F(y_f)$ of E_{U_S}/F, defined by $y_f^p - y_f = f$, is at most $(t-1)$ (see Equation (2.1)). Then, Theorem 2.3(iv) gives $g(E_{U_S}) \leq (t-1)(2^r - 1)$, which implies by Serre's bound that

$$|N(E_{U_S}) - (2^m + 1)| \leq \lfloor 2^{(m+2)/2} \rfloor (t-1)(2^r - 1). \qquad (4.7)$$

Since $f(P_0) = 0$, P_0 has 2 rational extensions in E_f (cf. Proposition 2.5). The place P_∞, on the other hand, is totally ramified in E_f/F and has a unique rational extension in E_f. Since these observations are valid for any $f \in U_S \subset V$, and using [50, Theorem III.8.4], P_0 has 2^r rational extensions in E_{U_S} and P_∞ has only one. Therefore, $|\overline{R}_{U_S}| = 2^r + 1$. Hence we have

$$\left| w(S) - 2^m(1 - 2^{-r}) \right| \leq (t-1)(1 - 2^{-r})\lfloor 2^{(m+2)/2} \rfloor. \qquad (4.8)$$

Since the above discussion holds for any r-dimensional subcode, we also obtain

$$d_r(C^\perp) \geq (1 - 2^{-r})\left(2^m - (t-1)\lfloor 2^{(m+2)/2} \rfloor \right). \qquad (4.9)$$

In [6, Theorem 4.2 and 4.3], the authors show that the above bounds are tight in some cases by exhibiting function fields which are maximal/minimal (cf. Section 5).

G. van der Geer and M. van der Vlugt have extensively studied trace codes and their generalized Hamming weights. Although there are also other contributions to the subject, we will cite some of their works since the methods used are parallel to the one described in this section.

- In [17], the following exact result on d_2 of the binary code $\mathrm{BCH}(2)^\perp$ of length $q - 1 = 2^m - 1$ is obtained for $m \geq 5$.

$$d_2(\mathrm{BCH}(2)^\perp) = \frac{3}{2}d_1(\mathrm{BCH}(2)^\perp) = \begin{cases} 3(q - \sqrt{2q})/4, & m \text{ odd} \\ 3(q - 2\sqrt{q})/4, & m \text{ even} \end{cases}$$

  Note that this result implies that the bound in Equation (4.9) is tight for $r = t = 2$ and m even.

- In [15], the authors compute the second and third generalized Hamming weights of binary Melas codes using properties of elliptic curves. They also obtain results on the second generalized weight of the dual code.

- We have seen that low weight codewords and low weight subcodes in trace codes correspond to Artin-Schreier and composita of Artin-Schreier function fields, respectively, with many rational places (cf. Equations (3.6) and (4.5)). In Sections 3 and 4, we used our knowledge on the number of rational places of function fields to estimate weights (higher weights) of trace codes. However, constructing function fields with many rational places is also an interesting problem (see the next section). In [20], the authors construct low weight codewords and subcodes of the generalized Reed-Muller codes to show the existence of function fields with many rational places. The same idea for other classes of codes can also be found in [18, 19].

5.   Maximal Function Fields

Let $F/\mathbb{F}_{q^2}$ be a function field with $\mathbb{F}_{q^2}$ as its full constant field. Using the Hasse-Weil bound for the number $N(F)$ of degree one places of F, we have

$$N(F) \leq q^2 + 1 + 2g(F)q,$$

where $g(F)$ is the genus of F. We call $F/\mathbb{F}_{q^2}$ *maximal* if

$$N(F) = q^2 + 1 + 2g(F)q.$$

There are many examples of maximal function fields of Artin-Schreier type. The most famous is the *Hermitian function field* $H = \mathbb{F}_{q^2}(x, y)$ which is defined by $y^q + y = x^{q+1}$. We can turn the defining equation of the Hermitian function field into the form in Definition 2.2. For this, let $\alpha \in \mathbb{F}_{q^2} \setminus \{0\}$ with $\alpha^q + \alpha = 0$, $y_1 = y/\alpha$ and $x_1 = x/\alpha$. Then

$$H = \mathbb{F}_{q^2}(x_1, y_1) \quad \text{with} \quad y_1^q - y_1 = \alpha x_1^{q+1}. \tag{5.1}$$

Using the affine model of H in (5.1), it is easy to determine the number of rational places of H. For each $t \in \mathbb{F}_{q^2}$ we have

$$\mathrm{Tr}_{\mathbb{F}_{q^2}/\mathbb{F}_q}(\alpha t^{q+1}) = \alpha t^{q+1} + \alpha^q t^{q+1} = 0$$

and hence the number of rational places of H is $q^3 + 1$. Using Equation (2.2) we obtain that the genus of H is $(q-1)q/2$, which implies that H is maximal.

For each $m \geq 2$ with $m|(q+1)$, the function field $F = \mathbb{F}_{q^2}(x, y)$ with $y^q + y = x^m$ is a subfield of H and hence is also maximal ([33]). For $\alpha \in \mathbb{F}_{q^2} \setminus \{0\}$ with $\alpha^q + \alpha = 0$, $y_1 = y/\alpha$ and $x_1 = x/\alpha$ we have

$$F = \mathbb{F}_{q^2}(x_1, y_1) \text{ with } y_1^q - y_1 = \alpha^{m-q} x_1^m.$$

Hence, F is also of Artin-Schreier type. Note that the genus of this function field is $(q-1)(m-1)/2$.

It is interesting to determine *possible genera* of maximal function fields, their *explicit equations* and to *classify* maximal function fields of a given genus [10]. We address these issues now.

A large class of maximum function fields are obtained using the Hermitian, Suzuki and Ree function fields, whose definitions we recall below.

- Hermitian function field: $\mathbb{F}_{q^2}(x, y)$ with

$$y^q + y = x^{q+1}.$$

- Suzuki function field: $\mathbb{F}_{2^{2s+1}}(x, y)$ with $s \geq 1$ and

$$y^{2^{2s+1}} + y = x^{2^s}\left(x^{2^{2s+1}} + x\right).$$

- Ree function field: $\mathbb{F}_{3^{2s+1}}(x, y, z)$ with $s \geq 1$ and

$$y^{3^{2s+1}} - y = x^{3^s}\left(x^{3^{2s+1}} - x\right), \quad z^{3^{2s+1}} - z = y^{2 \cdot 3^s}\left(y^{2s+1} - y\right).$$

All Hermitian function fields are maximal. For $m \geq 1$, the constant field extension $F\mathbb{F}_{2^{(2s+1)m}}$ of a Suzuki function field F is maximal if and only if $m \equiv 4 \mod 8$. Similarly for $m \geq 1$, the constant field extension $F\mathbb{F}_{3^{(2s+1) \cdot m}}$ of a Ree function field $F/\mathbb{F}_{3^{2s+1}}$ is maximal if and only if $m \equiv 6 \mod 12$ ([28], [42, Section 3]). We also note that new maximal function fields are obtained using subfields of maximal function fields [33], [12], [22], [3].

We have some restrictions on the genera of maximal function fields. It is known that for the genus g of a maximal function field over $\mathbb{F}_{q^2}$ we have [31]

$$g \leq q(q-1)/2.$$

Moreover there are also some gaps in the set $\{0, 1, \ldots, q(q-1)/2\}$ of nonnegative integers for the possible genera of maximal function fields over $\mathbb{F}_{q^2}$. For example it has been proved in [9, 52] that there is no maximal function field over $\mathbb{F}_{q^2}$ with

$$\left\lfloor \frac{(q-1)^2}{2} \right\rfloor + 1 \leq g \leq \frac{q(q-1)}{2} - 1.$$

We would like to state one of the first results in the classification of maximal function fields. In [44] it has been proved that the Hermitian function field over $\mathbb{F}_{q^2}$ is the unique maximal function field, up to isomorphism, over $\mathbb{F}_{q^2}$ with genus $q(q-1)/2$. There are further results in the classification of maximal function fields (cf. [8], [1])).

We denote the maximum number of degree one places among the function fields with full constant field $\mathbb{F}_q$ of genus g by $N_q(g)$. The Hasse-Weil bound immediately implies that $N_q(g) \leq q + 1 + 2g\sqrt{q}$. There are further improvements on $N_q(g)$ and we know the exact value of $N_q(g)$ only in some special cases, e.g. $g = 1$ [13]. We call that a function field of genus g with full constant field $\mathbb{F}_q$ is *optimal* if its number of degree one places is $N_q(g)$. From both theoretical and application point of view, it is of interest to construct function fields with as many rational places as possible (cf. [54], [50], [41]). We refer to [41] for various methods of such constructions. Applications also require explicit equations of such function fields. Let us give an example of an explicit function field which is constructed using Artin-Schreier extensions. We note that this is Example 4.4.7 in [41]. Let $F = \mathbb{F}_2(x, y_1, y_2)$ with

$$y_1^2 + y_1 = \frac{x(x_1)}{x^3 + x + 1}, \quad y_2^2 + y_2 = \frac{x(x+1)}{x^3 + x^2 + 1}.$$

The genus of F is 9 and its number of degree one places is 12. This function field over $\mathbb{F}_2$ is optimal.

For applications of function fields in the asymptotic theory of codes, it is important to find sequences of function fields $\{E_i\}_{i=1}^\infty$ with full constant field $\mathbb{F}_q$ such that $g(E_i) \to \infty$ and $\dfrac{N(E_i)}{g(E_i)} \to \lambda > 0$ as $i \to \infty$, where $N(E_i)$ and $g(E_i)$ are the number of rational places and the genus of E_i respectively (see also Chapter 1). The following result suggests that it may not be easy to find such a sequence of function fields.

Theorem 5.1 (Frey-Perret-Stichtenoth [7]). *Let $F/\mathbb{F}_q$ be an algebraic function field with $\mathbb{F}_q$ as its full constant field. Assume that $\{E_i\}_{i=1}^\infty$ is a sequence of abelian extensions of F such that $\mathbb{F}_q$ is the full constant field of E_i for $i \geq 1$. If $N(E_i)$ and $g(E_i)$ are the number of degree one places and the genus of E_i,* *for all i, and $lim_{i \to \infty} g(E_i) = \infty$, then*

$$\lim_{i \to \infty} \frac{N(E_i)}{g(E_i)} = 0.$$

In particular Theorem 5.1 implies that if E_i is an Artin-Schreier type extension over F for each $i \geq 1$ and $g(E_i) \to \infty$ as $i \to \infty$, then $\dfrac{N(E_i)}{g(E_i)} \to 0$.

References

[1] M. Abdon and F. Torres, "On maximal curves in characteristic two", Manuscripta Math., Vol. 99, 39-53 (1999).

[2] P. Charpin, "Open problems on cyclic codes", *Handbook of Coding Theory, North-Holland*, 963-1063 (1998).

[3] E. Çakçak and F. Özbudak, "Subfields of the function field of the Deligne-Luszting curve of Ree type", Acta Arith., Vol. 115, 133-180 (2004).

[4] P. Deligne, "Applications de la formule des traces aux sommes trigonométriques", *SGA $4\frac{1}{2}$ Cohomologie Etale*, Lecture Notes in Mathematics, Vol. 569, 168-232 (1978).

[5] P. Delsarte, "On Subfield Subcodes of Reed-Solomon Codes", IEEE Trans. Inform. Theory, Vol. 21, 575-576 (1975).

[6] I. Duursma, H. Stichtenoth and C. Voss, "Generalized Hamming weights for duals of BCH codes and maximal algebraic function fields", *Arithmetic, geometry and coding theory (Luminy, 1993)*, 53-65 (1996).

[7] G. Frey, M. Perret and H. Stichtenoth, "On the different of abelian extensions of global fields", *Coding Theory and Algebraic Geometry (Luminy 1991)*, Lecture Notes in Mathematics, Vol. 1518, 26-32 (1992).

[8] R. Fuhrman, A. Garcia and F. Torres, "On maximal curves", J. Number Theory, Vol. 67, 29-51 (1997).

[9] R. Fuhrman and F. Torres, "The genus of curves over finite fields with many rational points", Manuscripta Math., Vol. 89, 103-106 (1996).

[10] A. Garcia, "Curves over finite fields attaining the Hasse-Weil upper bound", *European Congress of Mathematics, Vol. II (Barcelona, 2000)*, Progr. Math., Vol. 202, 199-205 (2001).

[11] A. Garcia and H. Stichtenoth, "Elementary abelian p-extensions of algebraic function fields", Manuscripta Math., Vol. 72, 67-79 (1991).

[12] A. Garcia, H. Stichtenoth and C. P. Xing, "On subfields of the Hermitian function field", Compositio Math., Vol. 120, 137-170 (2000).

[13] G. van der Geer and M. van der Vlugt, Tables of curves with many points, available at http://www.science.uva.nl/~geer/tables-mathcomp15.ps.

[14] G. van der Geer and M. van der Vlugt, "Artin-Schreier curves and codes", J. Algebra, Vol. 139, 256-272 (1991).

[15] G. van der Geer and M. van der Vlugt, "Generalized Hamming weights of Melas codes and dual Melas codes", SIAM J. Discrete Math., Vol. 7, 554-559 (1994).

[16] G. van der Geer and M. van der Vlugt, "Fibre products of Artin-Schreier curves and generalized Hamming weights of codes", J. Combin. Theory Ser. A, Vol. 70, 337-348 (1995).

[17] G. van der Geer and M. van der Vlugt, "The second generalized Hamming weight of the dual codes of double-error correcting binary BCH codes", Bull. London Math. Soc., Vol. 27, 82-86 (1995).

[18] G. van der Geer and M. van der Vlugt, "Generalized Hamming weights of codes and curves over finite fields with many points", *Proceedings of the Hirzebruch 65 Conference on Algebraic Geometry (Ramat Gan, 1993)*, Israel Math. Conf. Proc., Vol. 9, 417-432 (1996).

[19] G. van der Geer and M. van der Vlugt, "Quadratic forms, generalized Hamming weights of codes and curves with many points", J. Number Theory, Vol. 59, 20-36 (1996).

[20] G. van der Geer and M. van der Vlugt, "Generalized Reed-Muller codes and curves with many points", J. Number Theory, Vol. 72, 257-268 (1998).

[21] G. van der Geer, R. Schoof and M. van der Vlugt, "Weight formulas for ternary Melas codes", Math. Comp., Vol. 58, 781-792 (1992).

[22] M. Giulietti, G. Korchmaros and F. Torres, "Quotient curves of the Deligne-Lusztig curve of Suzuki type", available at arXiv:math.AG/0206311, preprint, 2002.

[23] V.D. Goppa, "Codes on algebraic curves", Soviet Math. Dokl., Vol. 24, 170-172 (1981).

[24] C. Güneri, "Artin-Schreier curves and weights of two-dimensional cyclic codes", Finite Fields Appl., Vol. 10, 481-505 (2004).

[25] C. Güneri and F. Özbudak, "Improvements on generalized Hamming weights of some trace codes", Des. Codes Cryptogr, Vol. 39, 215-231 (2006).

[26] C. Güneri and F. Özbudak, "Multidimensional cyclic codes and Artin-Schreier hypersurfaces", preprint, 2006.

[27] C. Güneri and F. Özbudak, "Cyclic codes and reducible additive equations", preprint, 2006.

[28] J. P. Hansen and J. P. Pedersen, "Automorphism groups of Ree type, Deligne-Lusztig curves and function fields", J. Reine Angew. Math., Vol. 440, 99-109 (1993).

[29] H. Hasse, "Theorie der relativ zyklischen algebraischen Funktionenkörper", J. Reine Angew. Math., Vol. 172, 37-54 (1934).

[30] T. Helleseth, T. Kløve and Ø. Ytrehus "Generalized Hamming weights of linear codes", IEEE Trans. Inform. Theory, Vol. 38, 1133-1140 (1992).

[31] Y. Ihara, "Some remarks on the number of rational points of algebraic curves over finite fields", J. Fac. Sci. Tokio, Vol. 28, 721-724 (1981).

[32] T. Kløve, "Support weight distribution of linear codes", Discrete Math., Vol. 106/107, 311-316 (1992).

[33] G. Lachaud, "Sommes d'Eisenstein et nombre de points de certaines courbes algébriques sur les corps finis", C.R. Acad. Sci. Paris, Vol. 305, 729-732 (1987).

[34] S. Lang, *Algebra*, Addison-Wesley, 3rd ed., 1999.

[35] R. Lidl and H. Niederreieter, *Finite Fields*, Cambridge University Press, Cambridge, 1997.

[36] J.H. van Lint, *Introduction to Coding Theory*, Springer-Verlag, 1999.

[37] G. McGuire and J.F. Voloch, "Weights in codes and genus 2 curves", Proc. Amer. Math. Soc., Vol. 133, 2429-2437 (2005).

[38] O. Moreno and P.V. Kumar, "Minimum distance bounds for cyclic codes and Deligne's theorem", IEEE Trans. Inform. Theory, Vol. 39, 1524-1534 (1993).

[39] O. Moreno, J.P. Pedersen and D. Polemi, "An improved Serre bound for elementary abelian extensions of $\mathbb{F}_q(x)$ and the generalized Hamming weights of duals of BCH codes", IEEE Trans. Inform. Theory, Vol. 44, 1291-1293 (1998).

[40] F.J. MacWilliams and N.J.A. Sloane, *The theory of error-correcting codes*, North-Holland, 10th Ed., 1998.

[41] H. Niederreiter and C. Xing, *Rational Points on Curves over Finite Fields*, Cambridge University Press, Cambridge, 2001.

[42] J. P. Pedersen, "A function field related to the Ree group", *Coding Theory and Algebraic Geometry (Luminy 1991)*, Lecture Notes in Mathematics, Vol. 1518, 122-131 (1992).

[43] V. Pless, *Introduction to the Theory of Error-Correcting Codes*, Wiley, 1998.

[44] H. G. Rück and H. Stichtenoth, "A characterization of Hermitian function fields over finite fields", J. Reine Angew. Math., Vol. 457, 185-188 (1994).

[45] R. Schoof, "Nonsingular plane cubic curves over finite fields", J. Combin. Theory Ser. A, Vol. 46, 183-211 (1987).

[46] R. Schoof, "Families of curves and weight distributions of codes", Bull. Amer. Math. Soc., Vol. 32, 171-183 (1995).

[47] R. Schoof and M. van der Vlugt, "Hecke operators and the weight distributions of certain codes", J. Combin. Theory Ser. A, Vol. 57, 163-186 (1991).

[48] J. Simonis, "The effective length of subcodes", Appl. Algebra Engrg. Comm. Comput., Vol. 5, 371-377 (1994).

[49] H. Stichtenoth "On the dimension of subfield subcodes", IEEE Trans. Inform. Theory, Vol. 36, 90-93 (1990).

[50] H. Stichtenoth, *Algebraic Function Fields and Codes*, Springer-Verlag, Berlin, 1993.

[51] H. Stichtenoth and C. Voss, "Generalized Hamming weights of trace codes", IEEE Trans. Infrom. Theory, Vol. 40, 554-558 (1994).

[52] H. Stichtenoth and C. P. Xing, "The genus of maximal function fields over finite fields", Manuscripta Math., Vol. 86, 217-224 (1995).

[53] K-O. Stöhr and J.F. Voloch, "Weierstrass points and curves over finite fields", Proc. London Math. Soc., Vol. 52, 1-19 (1986).

[54] M. A. Tsfasman and S. G. Vladut, *Algebraic-Geometric Codes*, Kluwer, Dordrecht, 1991.

[55] M. A. Tsfasman and S. G. Vladut, "Geometric approach to higher weights", IEEE Trans. Inform. Theory, Vol. 41, 1564-1588 (1995).

[56] M. van der Vlugt, "The true dimension of certain binary Goppa codes", IEEE Trans. Inform. Theory, Vol. 36, 397-398 (1990).

[57] M. van der Vlugt, "On the dimension of trace codes", IEEE Trans. Inform. Theory, Vol. 37, 196-199 (1991).

[58] M. van der Vlugt, "A new upper bound for the dimension of trace codes", Bull. London Math. Soc., Vol. 23, 395-400 (1991).

[59] J.F. Voloch, "On the duals of binary BCH codes", IEEE Trans. Inform. Theory, Vol. 47, 2050-2051 (2001).

[60] V.K. Wei, "Generalized Hamming weights for linear codes", IEEE Trans. Inform. Theory, Vol. 37, 1412-1418 (1991).

[61] J. Wolfmann, "New bounds on cyclic codes from algebraic curves", Lecture Notes in Computer Science, Vol. 388, 47-62 (1989).

Chapter 4

PSEUDORANDOM SEQUENCES

Alev Topuzoğlu and Arne Winterhof

1.   Introduction

Sequences, which are generated by deterministic algorithms so as to simulate truly random sequences are said to be *pseudorandom* (PR). A pseudorandom sequence in the unit interval $[0, 1)$ is called a sequence of *pseudorandom numbers* (PRNs). In particular, for a prime p we represent the elements of the finite field $\mathbb{F}_p$ of p elements by the set $\{0, 1, ..., p - 1\}$, and arrive at a sequence of PRNs, say (y_n), through a sequence (x_n) over $\mathbb{F}_p$ satisfying $y_n = x_n/p$. The sequence (x_n) in this case is usually called a *pseudorandom number generator*.

Our main aim here is to elucidate the motivation for constructing PR sequences with some specific properties that foster their use in *cryptography* and in *quasi-Monte Carlo methods*. Our exposition focuses on some particular measures of "randomness" with respect to which "good" sequences have been constructed recently by the use of geometric methods. Some of these constructions are given in Chapter 2 of this book.

We also illustrate some typical methods that are used in the classical analysis of randomness of PRNs and briefly describe some recent approaches in order to familiarise the reader with basic notions and problems in this area of research. An extensive list of references is provided for the interested reader.

Various quality measures for randomness of PR sequences are in use. One should note here that the hierarchy among them varies according to the type of problem where PR sequences are needed. For example, if one wishes to employ a quasi-Monte Carlo method to approximate π by choosing N pairs $(x_n, x_{n+1}) \in [0, 1)^2, n = 0, 1, \ldots, N - 1$, of PRNs, counting the number K of

A. Garcia and H. Stichtenoth (eds.), *Topics in Geometry, Coding Theory and Cryptography*, 135–166.
© 2007 *Springer*.

pairs (x_n, x_{n+1}) in the unit circle and taking $\pi \approx 4K/N$, one should make sure that the PRNs in use are "distributed uniformly" in the unit square. On the other hand "unpredictability" is often the most desirable property for cryptographic applications, as it is described in Chapter 2, Section 3.

This chapter is structured as follows. We start with an outline of some basic facts regarding "linear complexity" and "linear complexity profile", which are potent measures of unpredictability (or, at least of predictability). Results on lower bounds for linear complexity and linear complexity profile for various PR sequences of wide interest are given in Section 2.1. We consider explicit and recursive nonlinear generators, in particular a new class of PR sequences, defined via Dickson polynomials and Rédei functions, and a generalisation of the well known inversive generator. Section 2.1 also deals with Legendre sequences and their variants, and the elliptic curve generators which have attracted considerable attention recently. In Section 2.2, we describe other measures related to linear complexity, with particular emphasis on the lattice test. In Sections 3 and 4 we turn our attention to measures of distribution; in particular we focus on autocorrelation and related concepts for binary sequences in Section 3. This may provide further background for Section 3 of Chapter 2. We conclude with Section 4 where we concentrate on discrepancy as a measure for uniform distribution of PRNs. Some recent results are presented which illustrate the well known relation of discrepancy to exponential sums. The significance of recent geometric constructions of low-discrepancy point sets is described. With the intention of keeping this chapter concise, we present primarily short or elementary proofs which are sufficiently indicative of some standard tools.

In the sequel we shall be concerned with PR sequences over a finite field $\mathbb{F}_q$ of $q = p^r$ elements with a positive integer r and a prime p. Note that a sequence (y_n) of PRNs in the unit interval can be obtained from a sequence (ξ_n) over $\mathbb{F}_q$ by $y_n = (k_r + k_{r-1}p + \ldots + k_1p^{r-1})/q$ if $\xi_n = k_1\beta_1 + \ldots + k_r\beta_r$ for some fixed ordered basis $\{\beta_1, \ldots, \beta_r\}$ of $\mathbb{F}_q$ over $\mathbb{F}_p$. A sequence over $\mathbb{F}_2$ is called a *bit sequence*. We shall restrict ourselves to (purely) periodic sequences, i.e., to those (ξ_n) satisfying $\xi_{n+t} = \xi_n$ for some positive integer t, for all $n \geq 0$.

We should note here that the term "pseudorandom number generator" is commonly used in the literature on pseudorandom sequences. In particular for sequences over $\mathbb{F}_p$ (identified with $\mathbb{Z}_p = \{0, 1, ..., p-1\}$) or the ring $\mathbb{Z}_m$, one refers to *congruential generators*. The "generator" here is sometimes used to mean the "(recurrence) relation" producing the sequence over $\mathbb{F}_p$ (or $\mathbb{Z}_m$), which in return gives rise to a PRN in the unit interval. In this Chapter we generally use the expression PR sequences. However the term congruential generator (or generator) will also appear when referring to some specific sequences over $\mathbb{F}_p$ (or $\mathbb{F}_q$), that are widely known as such.

2. Linear Complexity and Linear Complexity Profile

Linear complexity and linear complexity profile are defined in Chapter 2. We restate their definitions here for the convenience of the reader.

Let us first recall that a sequence $(s_n)_{n\geq 0}$ of elements of a ring R is called a *(homogeneous) linear recurring sequence of order k* if there exist elements $a_0, a_1, \ldots, a_{k-1}$ in R, satisfying the *linear recurrence of order k over R*;

$$s_{n+k} = a_{k-1}s_{n+k-1} + a_{k-2}s_{n+k-2} + \ldots + a_0 s_n, \quad n = 0, 1, \ldots$$

Now let (s_n) be a sequence over a ring R. One can associate to it a non-decreasing sequence $L(s_n, N)$ of non-negative integers as follows:

The *linear complexity profile* of a sequence (s_n) over the ring R is the sequence $L(s_n, N)$, $N \geq 1$, where its Nth term is defined to be the smallest L such that a linear recurrence of order L over R can generate the first N terms of (s_n).

We use the convention that $L(s_n, N) = 0$ if the first N elements of (s_n) are all zero and $L(s_n, N) = N$ if the first $N - 1$ elements of (s_n) are zero and $s_{N-1} \neq 0$.

The value

$$L(s_n) = \sup_{N \geq 1} L(s_n, N)$$

is called the *linear complexity* of the sequence (s_n). For the linear complexity of any periodic sequence of period t one easily verifies that

$$L(s_n) = L(s_n, 2t) \leq t.$$

Linear complexity and linear complexity profile of a given sequence (as well as the linear recurrence defining it) can be determined by using the well known Berlekamp-Massey algorithm (see e.g. [35]). The algorithm is efficient for sequences with low linear complexity and hence such sequences can easily be predicted. One typical example is the so-called "linear generator"

$$s_{n+1} = as_n + b, \tag{2.1}$$

for $a, b \in \mathbb{F}_p, a \neq 0$, with $L(s_n) \leq 2$. Faster algorithms are known for sequences of particular periods [26, 78, 79]. PR sequences with low linear complexity are shown to be unsuitable also for some applications using quasi-Monte Carlo methods (see [53, 55, 59]).

The expected values of linear complexity and linear complexity profile show that a "random" sequence should have $L(s_n, N)$ to be close to $\min\{N/2, t\}$ for all $N \geq 1$, see Chapter 2.

Two types of problems concerning linear complexity and linear complexity profile are of interest. One would like to construct sequences with high linear complexity (and possibly with other favourable properties). Chapter 2 illustrates such constructions. One would also like to find lower bounds for widely used PR sequences in order to judge whether it is reasonable to use them for cryptographic purposes. The present section focuses on this problem.

2.1 Lower Bounds for Linear Complexity and Linear Complexity Profile

Explicit Nonlinear Pseudorandom Sequences

It is possible to express linear complexity in connection with various invariants of the PR sequences at hand. The linear complexity profile of a sequence, for instance, can be determined utilising its generating function as described in Chapter 2.

In case of a q-periodic sequence (ξ_n) over $\mathbb{F}_q$, linear complexity is related to the degree of the polynomial $g \in \mathbb{F}_q[X]$ representing the sequence (ξ_n). We recall that the polynomial g can be uniquely determined as follows: Consider a fixed ordered basis $\{\beta_1, \ldots, \beta_r\}$ of $\mathbb{F}_q$ over $\mathbb{F}_p$, and for $n = n_1 + n_2 p + \ldots + n_r p^{r-1}$ with $0 \leq n_k < p, 1 \leq k \leq r$, order the elements of $\mathbb{F}_q$ as

$$\zeta_n = n_1\beta_1 + n_2\beta_2 + \ldots + n_r\beta_r.$$

Then g is the polynomial which satisfies $\deg g \leq q - 1$ and

$$\xi_n = g(\zeta_n), \quad 0 \leq n \leq q - 1. \tag{2.2}$$

When $\deg g \geq 2$, $q = p$ (and $\beta_1 = 1$) these sequences are called *explicit nonlinear congruential generators* and we have

$$L(\xi_n) = \deg g + 1 \tag{2.3}$$

(for a proof, see Blackburn *et al* [5, Theorem 8]). For a prime power q they are named *explicit nonlinear digital generators*. In general (2.3) is not valid for $r \geq 2$. Meidl and Winterhof [47] showed however that the following inequalities hold:

$$(\deg(g) + 1 + p - q)\frac{q}{p} \leq L(\xi_n) \leq (\deg(g) + 1)\frac{p}{q} + q - p.$$

For lower bounds on the linear complexity profile of (ξ_n) see Meidl and Winterhof [48].

A similar relation is valid for t-periodic sequences over $\mathbb{F}_q$ where t divides $q - 1$. For a t-periodic sequence (ω_n) one considers the unique polynomial $f \in \mathbb{F}_q[x]$ of degree at most $t - 1$, satisfying

$$\omega_n = f(\gamma^n), \quad n \geq 0,$$

for an element $\gamma \in \mathbb{F}_q$ of order t. In this case, $L(w_n)$ is equal to the number of nonzero coefficients of f (see [35]). Lower bounds for the linear complexity profile in some special cases are given by Meidl and Winterhof in [49]. For a general study of sequences with arbitrary periods see Massey and Serconek [42].

The following sequences exhibit a particularly nice behaviour with respect to the linear complexity profile. The *explicit inversive congruential generator* (z_n) was introduced by Eichenauer-Herrmann in [19]. The sequence (z_n) in this case is produced by the relation

$$z_n = (an + b)^{p-2}, \quad n = 0, \ldots, p-1, \quad z_{n+p} = z_n, \ n \geq 0, \qquad (2.4)$$

with $a, b \in \mathbb{F}_p$, $a \neq 0$, and $p \geq 5$. It is shown in [48] that

$$L(z_n, N) \geq \begin{cases} (N-1)/3, & 1 \leq N \leq (3p-7)/2, \\ N - p + 2, & (3p-5)/2 \leq N \leq 2p - 3, \\ p - 1, & N \geq 2p - 2. \end{cases} \qquad (2.5)$$

We provide the proof of a slightly weaker result.

Theorem 2.1. *Let (z_n) be as in (2.4), then*

$$L(z_n, N) \geq \min\left\{\frac{N-1}{3}, \frac{p-1}{2}\right\}, \quad N \geq 1.$$

Proof. Suppose (z_n) satisfies a linear recurrence relation of length L,

$$z_{n+L} = c_{L-1}z_{n+L-1} + \ldots + c_0 z_n, \quad 0 \leq n \leq N - L - 1, \qquad (2.6)$$

with $c_0, \ldots, c_{L-1} \in \mathbb{F}_p$. We may assume $L \leq p - 1$. Put

$$C_L(N) = \{n; 0 \leq n \leq \min\{N - L, p\} - 1, a(n+l) + b \neq 0, 0 \leq l \leq L\}$$

Note that $\#C_L(N) \geq \min\{p, N - L\} - (L + 1)$.

For $n \in C_L(N)$ the recurrence (2.6) is equivalent to

$$(a(n + L) + b)^{-1} = c_{L-1}(a(n + L - 1) + b)^{-1} + \ldots + c_0(an + b)^{-1}.$$

Multiplication with

$$\prod_{j=0}^{L}(a(n + j) + b)$$

yields

$$\prod_{j=0}^{L-1}(a(n + j) + b) = \sum_{l=0}^{L-1} c_l \prod_{\substack{j=0 \\ j \neq l}}^{L}(a(n + j) + b)$$

for all $n \in C_L(N)$. Hence the polynomial

$$F(X) = -\prod_{j=0}^{L-1}(a(X+j)+b) + \sum_{l=0}^{L-1} c_l \prod_{\substack{j=0 \\ j \neq l}}^{L}(a(X+j)+b)$$

is of degree at most L and has at least $\min\{p, N-L\} - (L+1)$ zeros. On the other hand

$$F(-a^{-1}b - L) = -a^L \prod_{j=0}^{L-1}(j-L) \neq 0,$$

hence $F(X)$ is not the zero polynomial and we get

$$L \geq \deg(F) \geq \min\{p, N-L\} - (L+1),$$

which implies the desired result.   □

Analogues of (2.5) for *digital inversive generators*, i.e., for $r \geq 2$, are also given in [48]. For *t-periodic inversive generators*, where t is a divisor of $q-1$, see [49].

We mention one more explicit nonlinear generator, namely the *quadratic exponential generator*, introduced by Gutierrez *et al* [32]. Given an element $\vartheta \in \mathbb{F}_q^*$ we consider the sequence (q_n) where

$$q_n = \vartheta^{n^2}, \qquad n = 0, 1, \ldots .$$

The lower bound

$$L(q_n, N) \geq \frac{\min\{N, t\}}{2}, \qquad N \geq 1,$$

is obtained in [32]. Here t is at least $\tau/2$ where τ is the multiplicative order of ϑ.

Recursive Nonlinear Pseudorandom Sequences

Given a polynomial $f(X) \in \mathbb{F}_p[X]$ of degree $d \geq 2$, the *nonlinear congruential pseudorandom number generator* (u_n) is defined by the recurrence relation

$$u_{n+1} = f(u_n), \quad n \geq 0, \tag{2.7}$$

with some initial value $u_0 \in \mathbb{F}_p$. Obviously, the sequence (u_n) is eventually periodic with some period $t \leq p$. As usual, we assume it to be purely periodic.

The following lower bound on the linear complexity profile of a nonlinear congruential generator is given in [32].

Theorem 2.2. *Let (u_n) be as in (2.7), where $f(X) \in \mathbb{F}_p[X]$ is of degree $d \geq 2$, then*

$$L(u_n, N) \geq \min \{\log_d(N - \lfloor \log_d N \rfloor), \log_d t\}, \quad N \geq 1.$$

Proof. Let us consider the following sequence of polynomials over $\mathbb{F}_p$:

$$F_0(X) = X, \qquad F_i(X) = F_{i-1}(f(X)), \qquad i = 1, 2, \ldots.$$

It is clear that $\deg(F_i) = d^i$ for every $i = 1, 2, \ldots$. Moreover $u_{n+j} = F_j(u_n)$ for any integers $n, j \geq 0$. Put $L = L(u_n, N)$ so that we have

$$u_{n+L} = \sum_{l=0}^{L-1} c_l u_{n+l}, \qquad 0 \leq n \leq N - L - 1,$$

for some $c_0, \ldots, c_{L-1} \in \mathbb{F}_p$. Therefore the polynomial

$$F(X) = -F_L(X) + \sum_{l=0}^{L-1} c_l F_l(X)$$

has degree d^L and at least $\min \{N - L, t\}$ zeros. Thus $d^L \geq \min \{N - L, t\}$. Since otherwise the result is trivial, we may suppose $L \leq \lfloor \log_d N \rfloor$ and get $d^L \geq \min \{N - \lfloor \log_d N \rfloor, t\}$, which yields the assertion. $\square$

For some special classes of polynomials much better results are available, see [30, 32, 65]. For instance, in case of the largest possible period $t = p$ we have

$$L(u_n, N) \geq \min\{N - p + 1, p/d\}, \quad N \geq 1.$$

The *inversive (congruential) generator* (y_n) defined by

$$y_{n+1} = a y_n^{p-2} + b = \begin{cases} a y_n^{-1} + b & \text{if } y_n \neq 0, \\ b & \text{otherwise,} \end{cases} \quad n \geq 0, \qquad (2.8)$$

with $a, b, y_0 \in \mathbb{F}_p$, $a \neq 0$, has linear complexity profile

$$L(y_n, N) \geq \min \left\{ \frac{N-1}{3}, \frac{t-1}{2} \right\}, \quad N \geq 1. \qquad (2.9)$$

This sequence, introduced by Eichenauer and Lehn [18], has succeeded in drawing significant attention due to some of its enchanting properties. In terms of the linear complexity profile the lower bound (2.9) shows that the inversive generator is almost optimal. This aspect will be reconsidered in Section 2.2. The sequence (y_n) attains the largest possible period $t = p$ if, for instance, $X^2 - aX - b$ is a primitive polynomial over $\mathbb{F}_p$. See Flahive and Niederreiter [21] for a refinement of this result.

The *power generator* (p_n), defined as

$$p_{n+1} = p_n^e, \quad n \geq 0,$$

with some integer $e \geq 2$ and initial value $0 \neq p_0 \in \mathbb{F}_p$ satisfies

$$L(p_n, N) \geq \min \left\{ \frac{N^2}{4(p-1)}, \frac{t^2}{p-1} \right\}, \quad N \geq 1.$$

Results about the period length of (p_n) can be found in Friedlander *et al* [23, 24].

The family of *Dickson polynomials* $D_e(X, a) \in \mathbb{F}_p[X]$ is defined by the recurrence relation

$$D_e(X, a) = X D_{e-1}(X, a) - a D_{e-2}(X, a), \quad e = 2, 3, \ldots,$$

with initial values $D_0(X, a) = 2$, $D_1(X, a) = X$, where $a \in \mathbb{F}_p$. Obviously, the degree of D_e is e. It is easy to see that $D_e(X, 0) = X^e$, $e \geq 2$, which corresponds to the case of the power generator. In the special case that $a = 1$ the lower bound

$$L(u_n, N) \geq \frac{\min\{N^2, 4t^2\}}{16(p+1)} - (p+1)^{1/2}, \quad N \geq 1,$$

for a new class of nonlinear congruential generators where $f(X) = D_e(X, 1)$ is proven by Aly and Winterhof [1]. Here t is a divisor of $p - 1$ or $p + 1$.

Another class of nonlinear congruential pseudorandom number generators, where $f(X)$ is a Rédei function, is analysed by Meidl and Winterhof [52]. Suppose that

$$r(X) = X^2 - \alpha X - \beta \in \mathbb{F}_p[X]$$

is an irreducible quadratic polynomial with the two different roots ξ and $\zeta = \xi^p$ in $\mathbb{F}_{p^2}$. We consider the polynomials $g_e(X)$ and $h_e(X) \in \mathbb{F}_p[X]$, which are uniquely defined by the equation

$$(X + \xi)^e = g_e(X) + h_e(X)\xi.$$

The *Rédei function* $f_e(X)$ of degree e is then given by

$$f_e(X) = \frac{g_e(X)}{h_e(X)}.$$

The function $f_e(X)$ is a permutation of $\mathbb{F}_p$ if and only if $\gcd(e, p+1) = 1$, see Nöbauer [63]. For further background on Rédei functions we refer to [41, 63]. We consider generators (r_n) defined by

$$r_{n+1} = f_e(r_n), \quad n \geq 0,$$

with a Rédei permutation $f_e(X)$ and some initial element $u_0 \in \mathbb{F}_p$. The sequence (r_n) is periodic with period t, where t is a divisor of $\varphi(p+1)$ and φ is the Euler φ-function. As any mapping over $\mathbb{F}_p$, the Rédei permutation can be uniquely represented by a polynomial of degree at most $p-1$ and therefore the sequence (r_n) belongs to the class of nonlinear congruential pseudorandom number generators (2.7). In [52] the following lower bound on the linear complexity profile of the sequence (r_n) is obtained:

$$L(r_n, N) \geq \frac{\min\{N^2, 4t^2\}}{20(p+1)^{3/2}}, \quad N \geq 2,$$

provided that $t \geq 2$.

The linear complexity profile of pseudorandom number generators over $\mathbb{F}_p$, defined by a recurrence relation of order $m \geq 1$ is studied in Topuzoğlu and Winterhof [71];

$$u_{n+1} = f(u_n, u_{n-1}, \ldots, u_{n-m+1}), \; n = m-1, m, \ldots. \tag{2.10}$$

Here initial values $u_0, \ldots, u_{m-1}$ are in $\mathbb{F}_p$ and $f \in \mathbb{F}_p(X_1, \ldots, X_m)$ is a rational function in m variables over $\mathbb{F}_p$. The sequence (2.10) eventually becomes periodic with least period $t \leq p^m$. The fact that t can actually attain the value p^m gains nonlinear generators of higher orders a particular interest. In case of a polynomial f, lower bounds for the linear complexity and linear complexity profile of higher order generators are given in [71].

A particular rational function f in (2.10) gives rise to a generalisation of the inversive generator (2.8), as described below. Let (x_n) be the sequence over $\mathbb{F}_p$, defined by the linear recurrence relation of order $m+1$;

$$x_{n+1} = a_0 x_n + a_1 x_{n-1} + \ldots + a_m x_{n-m}, \quad n \geq m,$$

with $a_0, a_1, \ldots, a_m \in \mathbb{F}_p$ and initial values $x_0, \ldots, x_m \in \mathbb{F}_p$. An increasing function $N(n)$ is defined by

$$N(0) = \min\{n \geq 0 : x_n \neq 0\},$$
$$N(n) = \min\{l \geq N(n-1) + 1 : x_l \neq 0\},$$

and the nonlinear generator (z_n) is produced by

$$z_n = x_{N(n)+1} x_{N(n)}^{-1}, \quad n \geq 0$$

(see Eichenauer et.al. [17]). It is easy to see that (z_n) satisfies

$$z_{n+1} = f(z_n, \ldots, z_{n-m+1}), \quad n \geq m-1,$$

whenever $z_n \cdots z_{n-m+1} \neq 0$ for the rational function

$$f(X_1, \ldots, X_m) = a_0 + a_1 X_1^{-1} + a_2 X_1^{-1} X_2^{-1} + \ldots + a_m X_1^{-1} X_2^{-1} \cdots X_m^{-1}.$$

A sufficient condition for (z_n) to attain the maximal period length p^m is given in [17]. It is shown in [71] that the linear complexity profile $L(z_n, N)$ of (z_n) with the least period p^m satisfies

$$L(z_n, N) \geq \min\left(\left\lceil \frac{p-m}{m+1} \right\rceil p^{m-1} + 1, N - p^m + 1 \right), \quad N \geq 1.$$

This result is in accordance with (2.9), i.e., the case $m = 1$.

Legendre Sequence and Related Bit Sequences

Let $p > 2$ be a prime. The *Legendre-sequence* (l_n) is defined by

$$l_n = \begin{cases} 1, & \left(\frac{n}{p}\right) = -1, \\ 0, & \text{otherwise,} \end{cases} \quad n \geq 0,$$

where $\left(\frac{\cdot}{p}\right)$ is the Legendre-symbol. Obviously, (l_n) is p-periodic. Results on the linear complexity of (l_n) can be found in [13, 73]. We give the proof here since the method is illustrative.

Theorem 2.3. *The linear complexity of the Legendre sequence is*

$$L(l_n) = \begin{cases} (p-1)/2, & p \equiv 1 \bmod 8, \\ p, & p \equiv 3 \bmod 8, \\ p - 1, & p \equiv 5 \bmod 8, \\ (p+1)/2, & p \equiv 7 \bmod 8. \end{cases}$$

Proof. We start with the well known relation

$$L(l_n) = p - \deg(\gcd(S(X), X^p - 1))$$

where

$$S(X) = \sum_{n=0}^{p-1} l_n X^n,$$

(see for example [66, Lemma 8.2.1]), i.e., in order to determine the linear complexity it is sufficient to count the number of common zeros of $S(X)$ and $X^p - 1$ in the splitting field $\mathbb{F}$ of $X^p - 1$ over $\mathbb{F}_2$. Let $1 \neq \beta \in \mathbb{F}$ be a root of $X^p - 1$. For q with $\left(\frac{q}{p}\right) = 1$ we have

$$S(\beta^q) = \sum_{n=0}^{p-1} l_n \beta^{nq} = \sum_{\left(\frac{n}{p}\right)=-1} \beta^{nq} = \sum_{\left(\frac{n}{p}\right)=-1} \beta^n = S(\beta)$$

and for m with $\left(\frac{m}{p}\right) = -1$,

$$
\begin{aligned}
S(\beta^m) &= \sum_{\left(\frac{n}{p}\right)=-1} \beta^{nm} = \sum_{\left(\frac{n}{p}\right)=1} \beta^n \\
&= \sum_{n=1}^{p-1}(1+l_n)\beta^n = \frac{\beta^p - \beta}{\beta - 1} + S(\beta) = 1 + S(\beta).
\end{aligned}
$$

Moreover, we have $S(\beta) \in \mathbb{F}_2$ if and only if $S(\beta)^2 = S(\beta^2) = S(\beta)$, i.e., $\left(\frac{2}{p}\right) = 1$ which is equivalent to $p \equiv \pm 1 \bmod 8$. Next we have

$$
S(1) = \sum_{\left(\frac{n}{p}\right)=-1} 1 = \frac{p-1}{2} = \begin{cases} 0 & \text{if } p \equiv 1 \bmod 4, \\ 1 & \text{if } p \equiv 3 \bmod 4. \end{cases}
$$

Let Q and N denote the sets of quadratic residues and nonresidues modulo p, respectively. If $p \equiv \pm 1 \bmod 8$ then we have one of the following two cases: Either $S(\beta^q) = S(\beta^m) + 1 = 0$ for all $q \in Q$ and $m \in N$, or $S(\beta^m) = S(\beta^q) + 1 = 0$ for all $q \in Q$ and $m \in N$. Now the assertion is clear since $|Q| = |N| = (p-1)/2$. $\qquad\square$

The profile can be estimated by using bounds on incomplete sums of Legendre symbols. The proof below essentially follows that of [66, Theorem 9.2].

Theorem 2.4. *The linear complexity profile of the Legendre sequence satisfies*

$$
L(l_n, N) > \frac{\min\{N, p\}}{1 + p^{1/2}(1 + \log p)} - 1, \quad N \geq 1.
$$

Proof. Since $L(l_n, N) \geq L(l_n, p)$ for $N > p$ we may assume $N \leq p$. As usual, put $L = L(l_n, N)$ so that

$$
l_{n+L} = c_{L-1}l_{n+L-1} + \ldots + c_0 l_n, \quad 0 \leq n \leq N - L - 1,
$$

for some $c_0, \ldots, c_{L-1} \in \mathbb{F}_2$. Since $(-1)^{l_n} = \left(\frac{n}{p}\right), 1 \leq n \leq p-1$, with $c_L = 1$ we have

$$
1 = (-1)^{\sum_{j=0}^{L} c_j l_{n+j}} = \left(\frac{\prod_{j=0}^{L}(n+j)^{c_j}}{p}\right), \quad 1 \leq n \leq N - L - 1,
$$

and thus

$$
N - L - 1 = \sum_{n=1}^{N-L-1} \left(\frac{\prod_{j=0}^{L}(n+j)^{c_j}}{p}\right).
$$

The following bound for the right hand side of this equation

$$\left| \sum_{n=1}^{N-L-1} \left(\frac{\prod_{j=0}^{L}(n+j)^{c_j}}{p} \right) \right| < (L+1)p^{1/2}(1+\log p) \qquad (2.11)$$

yields

$$N - (L+1) < (L+1)p^{1/2}(1+\log p)$$

from which the assertion follows. The bound (2.11) can be proved as follows: For an integer $k \geq 2$ put $e_k(x) = \exp(2\pi i x/k)$. The relations below can be found in [74];

$$\sum_{a=0}^{k-1} e_k(au) = \begin{cases} 0, & u \not\equiv 0 \bmod k, \\ k, & u \equiv 0 \bmod k, \end{cases} \qquad (2.12)$$

$$\sum_{a=1}^{k-1} \left| \sum_{x=0}^{K-1} e_k(ax) \right| \leq k \log k, \quad 1 \leq K \leq k. \qquad (2.13)$$

The Weil bound, which we present in the following form (see [64, Theorems 2C and 2G]),

$$\left| \sum_{a=0}^{p-1} \chi(f(a))e_p(ax) \right| \leq \begin{cases} p^{1/2} \deg f, & 1 \leq x < p, \\ p^{1/2}(\deg f - 1), & x = 0, \end{cases} \qquad (2.14)$$

where χ denotes a nontrivial multiplicative character of $\mathbb{F}_p$ and $f \in \mathbb{F}_p[X]$, enables us to handle the complete hybrid character sum below. Application of Vinogradov's method (see [70]) with (2.12) and

$$f(X) = \prod_{j=0}^{L}(X+j)^{c_j}$$

gives

$$\left| \sum_{n=1}^{N-L-1} \left(\frac{f(n)}{p} \right) \right| = \frac{1}{p} \left| \sum_{x \in \mathbb{F}_p} \sum_{m \in \mathbb{F}_p} \left(\frac{f(m)}{p} \right) \sum_{n=1}^{N-L-1} e_p(x(n-m)) \right|$$

$$\leq \frac{1}{p} \sum_{x \in \mathbb{F}_p} \left| \sum_{m \in \mathbb{F}_p} \left(\frac{f(m)e_p(-xm)}{p} \right) \right| \left| \sum_{n=1}^{N-L-1} e_p(xn) \right|$$

$$< (L+1)p^{1/2}(1+\log p),$$

where we used that f is not a square (since $c_L = 1$) to apply (2.14) in the case $x = 0$. $\qquad \square$

For similar sequences, that are defined by the use of the quadratic character of arbitrary finite fields and the study of their linear complexity profiles, see [39, 46, 76].

Let γ be a primitive element and η be the quadratic character of the finite field $\mathbb{F}_q$ of odd characteristic. The *Sidelnikov sequence* (σ_n) is defined by

$$\sigma_n = \begin{cases} 1, & \text{if } \eta(\gamma^n + 1) = -1, \\ 0, & \text{otherwise}, \end{cases} \qquad n \geq 0.$$

In many cases one is able to determine the linear complexity $L(\sigma_n)$ over $\mathbb{F}_2$ exactly, see Meidl and Winterhof [51]. For example, if $(q-1)/2$ is an odd prime such that 2 is a primitive root modulo $(q-1)/2$, then (s_n) attains the largest possible linear complexity $L(\sigma_n) = q-1$. Moreover we have the lower bound, see [51],

$$L(\sigma_n, N) = \Omega\left(\frac{\min\{N, q\}}{q^{1/2} \log q}\right), \qquad N \geq 1.$$

The linear complexity over $\mathbb{F}_p$ of this sequence has been estimated in Garaev *et al* [27] by using bounds of character sums with middle binomial coefficients. For small values of p the linear complexity can be evaluated explicitly.

Let p and q be two distinct odd primes. Put

$$Q = \{q, 2q, \ldots, (p-1)q\}, \ Q_0 = Q \cup \{0\},$$

and

$$P = \{p, 2p, \ldots, (q-1)p\}.$$

The pq-periodic sequence (t_n) over $\mathbb{F}_2$, defined by

$$t_n = \begin{cases} 0, & \text{if } (n \bmod pq) \in Q_0, \\ 1, & \text{if } (n \bmod pq) \in P, \\ \left(1 - \left(\frac{n}{p}\right)\left(\frac{n}{q}\right)\right)/2, & \text{otherwise} \end{cases}$$

is called the *two-prime generator* (or *generalised cyclotomic sequence of order 2*) (see [10], and [13, Chapter 8.2]). Under the restriction $\gcd(p-1, q-1) = 2$

it satisfies

$$L(t_n) = \begin{cases} pq - 1, & p \equiv 1 \bmod 8 \text{ and } q \equiv 3 \bmod 8 \\ & \text{or } p \equiv 5 \bmod 8 \text{ and } q \equiv 7 \bmod 8, \\ (p-1)q, & p \equiv 7 \bmod 8 \text{ and } q \equiv 3 \bmod 8 \\ & \text{or } p \equiv 3 \bmod 8 \text{ and } q \equiv 7 \bmod 8, \\ pq - p - q + 1, & p \equiv 7 \bmod 8 \text{ and } q \equiv 5 \bmod 8 \\ & \text{or } p \equiv 3 \bmod 8 \text{ and } q \equiv 1 \bmod 8, \\ (pq + p + q - 3)/2, & p \equiv 1 \bmod 8 \text{ and } q \equiv 7 \bmod 8 \\ & \text{or } p \equiv 5 \bmod 8 \text{ and } q \equiv 3 \bmod 8, \\ (p-1)(q-1)/2, & p \equiv 7 \bmod 8 \text{ and } q \equiv 1 \bmod 8 \\ & \text{or } p \equiv 3 \bmod 8 \text{ and } q \equiv 5 \bmod 8, \\ (p-1)(q+1)/2, & p \equiv 7 \bmod 8 \text{ and } q \equiv 7 \bmod 8 \\ & \text{or } p \equiv 3 \bmod 8 \text{ and } q \equiv 3 \bmod 8. \end{cases}$$

In the most important case when $|p - q|$ is small we have a lower bound on the linear complexity profile of order of magnitude

$$O(N^{1/2}(pq)^{-1/4} \log^{-1/2}(pq))$$

for $2 \le N < pq$.

Elliptic Curve Generators

We recall some definitions and basic facts about elliptic curves (see [37] or Chapter 5).

Let $p > 3$ be a prime and E be an elliptic curve over $\mathbb{F}_p$ of the form

$$Y^2 = X^3 + aX + b$$

where the coefficients a, b are in $\mathbb{F}_p$ and $4a^3 + 27b^2 \neq 0$. The set $E(\mathbb{F}_p)$ of all $\mathbb{F}_p$-rational points on E forms an abelian group where we denote addition by $\oplus$. The point O at infinity is the zero element of $E(\mathbb{F}_p)$. We recall the Hasse-Weil bound

$$|\#E(\mathbb{F}_p) - p - 1| \le 2p^{1/2},$$

where $\#E(\mathbb{F}_p)$ is the number of $\mathbb{F}_p$-rational points, including O. For a given initial value $W_0 \in E(\mathbb{F}_p)$, a fixed point $G \in E(\mathbb{F}_p)$ of order t and a rational function $f \in \mathbb{F}_p(E)$ the *elliptic curve congruential generator* (with respect to f) is defined by $w_n = f(W_n)$, $n \ge 0$, where

$$W_n = G \oplus W_{n-1} = nG \oplus W_0, \quad n \ge 1.$$

Obviously, (w_n) is t-periodic. See [3, 34] and references therein for results on the properties of elliptic curve generators. For example, choosing the function

$f(x, y) = x$, the work of Hess and Shparlinski [34] gives the following lower bound for the linear complexity profile:

$$L(w_n, N) \geq \min\{N/3, t/2\}, \quad N \geq 2.$$

Here we present an elementary proof of a slightly weaker result. Let $x(Q)$ denote the first coordinate x of the point $Q = (x, y) \in E$.

Theorem 2.5. *Let (w_n) be the t-periodic sequence defined by*

$$w_n = x(nG), \quad 1 \leq n \leq t - 1, \tag{2.15}$$

with some $w_0 \in \mathbb{F}_p$ and $G \in E$ of order t. Then we have

$$L(w_n, N) \geq \frac{\min\{N, t/2\} - 3}{4}, \quad N \geq 2.$$

Proof. We may assume $N \leq t/2$ and $L(w_n, N) < t/2$. Put $nG = (x_n, y_n)$, $1 \leq n \leq t - 1$. Note that $x_k = x_m$ if and only if $k = m$ or $k = t - m$, $1 \leq k \leq t - 1$, and $y_k = 0$ if and only if t is even and $k = t/2$. Put $c_L = -1$ and assume that

$$\sum_{l=0}^{L} c_l w_{n+l} = 0, \quad L + 1 \leq n \leq N - L - 1,$$

or equivalently

$$\sum_{l=0}^{L} c_l w_{t-n-l} = 0, \quad L + 1 \leq n \leq N - L - 1.$$

Hence,

$$\sum_{l=0}^{L} c_l \frac{w_{n+l} + w_{t-n-l}}{2} = 0, \quad L + 1 \leq n \leq N - L - 1.$$

By the addition formulas for points on elliptic curves we have

$$
\begin{aligned}
x_{n+l} &= \left(\frac{y_n - y_l}{x_n - x_l}\right)^2 - (x_n + x_l) \\
&= \frac{x_l x_n^2 + (x_l^2 + a)x_n + ax_l + 2b - 2y_l y_n}{(x_n - x_l)^2}, \quad l + 1 \leq n \leq t - l - 1,
\end{aligned}
$$

where we used $y_n^2 = x_n^3 + ax_n + b$. Similarly we get

$$x_{t-n-l} = \frac{x_l x_n^2 + (x_l^2 + a)x_n + ax_l + 2b + 2y_l y_n}{(x_n - x_l)^2}, \quad l + 1 \leq n \leq t - l - 1,$$

and hence

$$\frac{x_{n+l} + x_{t-n-l}}{2} = \frac{x_l x_n^2 + (x_l^2 + a)x_n + ax_l + 2b}{(x_n - x_l)^2}, \quad l+1 \leq n \leq t-l-1.$$

So we get

$$\sum_{l=0}^{L} c_l \frac{x_l x_n^2 + (x_l^2 + a)x_n + ax_l + 2b}{(x_n - x_l)^2} = 0, \quad L+1 \leq n \leq N-L-1.$$

Clearing denominators we get

$$\sum_{l=0}^{L} c_l(x_l x_n^2 + (x_l^2+a)x_n + ax_l + 2b) \prod_{\substack{j=0 \\ j\neq l}}^{L} (x_n - x_j)^2 = 0, \quad L+1 \leq n \leq N-L-1.$$

So the polynomial

$$F(X) = \sum_{l=0}^{L} c_l(x_l X^2 + (x_l^2 + a)X + ax_l + 2b) \prod_{\substack{j=0 \\ j\neq l}}^{L} (X - x_j)^2$$

of degree at most $2(L+1)$ has at least $N - 2L - 1$ different zeros. Moreover, we have

$$F(x_L) = -2(x_L^3 + ax_L + b) \prod_{j=0}^{L-1} (x_L - x_j)^2 = -2y_L^2 \prod_{j=0}^{L-1} (x_L - x_j)^2 \neq 0.$$

Hence we get $2(L+1) \geq N - 2L - 1$ and the result follows.   □

2.2   Related Measures

Lattice Test

In order to study the structural properties of a given periodic sequence (s_n) over $\mathbb{F}_q$, it is natural to consider the subspaces $\mathcal{L}(s_n, s)$ of $\mathbb{F}_q^s$ for $s \geq 1$, spanned by the vectors $\mathbf{s}_n - \mathbf{s}_0$, $n = 1, 2, \ldots$ where

$$\mathbf{s}_n = (s_n, s_{n+1}, \ldots, s_{n+s-1}), \quad n = 0, 1, \ldots.$$

We recall that (s_n) is said to pass the *s-dimensional lattice test* for some integer $s \geq 1$, if $\mathcal{L}(s_n, s) = \mathbb{F}_q^s$. It is obvious for example that the linear generator (2.1) can pass the s-dimensional lattice test at most for $s = 1$. On the other hand for $q = p$, the nonlinear generator (2.2) passes the test for all $s \leq \deg g$ (see [53]). However this test is well known to be unreliable since sequences, which pass the lattice test for large dimensions, yet having bad statistical properties are known [53].

Accordingly the notion of *lattice profile* is introduced by Dorfer and Winterhof [16]. For given $s \geq 1$ and $N \geq 2$ we say that (s_n) passes the *s-dimensional N-lattice test* if the subspace spanned by the vectors $\mathbf{s}_n - \mathbf{s}_0$, $1 \leq n \leq N - s$, is $\mathbb{F}_q^s$. The largest s for which (s_n) passes the s-dimensional N-lattice test is called the *lattice profile at N*, and is denoted by $S(s_n, N)$.

The lattice profile is closely related to the linear complexity profile, as the following result in [16] shows:

We have either

$$S(s_n, N) = \min\{L(s_n, N), N + 1 - L(s_n, N)\}$$

or

$$S(s_n, N) = \min\{L(s_n, N), N + 1 - L(s_n, N)\} - 1.$$

The results of Dorfer *et al* [15] on the expected value of the lattice profile show that a "random" sequence should have $S(s_n, N)$ to be close to $\min\{N/2, t\}$.

k-Error Linear Complexity

We have remarked that a cryptographically strong sequence necessarily has a high linear complexity. It is also clear that the linear complexity of such a sequence should not decrease significantly when a small number of its terms are altered. The error linear complexity is introduced in connection with this observation [14, 69].

Let (s_n) be a sequence over $\mathbb{F}_q$, with period t. The *k-error linear complexity* $L_k(s_n)$ of (s_n) is defined as

$$L_k(s_n) = \min_{(y_n)} L(y_n),$$

where the minimum is taken over all t-periodic sequences (y_n) over $\mathbb{F}_q$, for which the Hamming distance of the vectors $(s_0, \ldots, s_{t-1})$ and $(y_0, \ldots, y_{t-1})$ is at most k.

One problem of interest here is to determine the minimum value k, for which $L_k(s_n) < L(s_n)$. This problem is tackled by Meidl [44], in case (s_n) is a bit sequence with period length p^n, where p is an odd prime and 2 is a primitive root modulo p^2. Meidl [44] also describes an algorithm to determine the k-error linear complexity that is based on an algorithm of [79]. Stronger results for p^n-periodic sequences over $\mathbb{F}_p$ have been recently obtained in Meidl [45].

Here we give the proof of the following recent result on the k-error linear complexity over $\mathbb{F}_p$ of Legendre sequences, obtained by Aly and Winterhof in [2].

Theorem 2.6. *Let $L_k(l_n)$ denote the k-error linear complexity over $\mathbb{F}_p$ of the Legendre sequence (l_n). Then,*

$$L_k(l_n) = \begin{cases} p, & k = 0, \\ (p+1)/2, & 1 \le k \le (p-3)/2, \\ 0, & k \ge (p-1)/2. \end{cases}$$

Proof. Put

$$g_1(X) = \frac{1}{2}\left(X^{p-1} - X^{(p-1)/2}\right) \text{ and } g_2(X) = \frac{1}{2}\left(1 - X^{(p-1)/2}\right).$$

Since $l_n = g_1(n)$ for $n \ge 0$ we get that the Legendre sequence (l_n) over $\mathbb{F}_p$ has linear complexity $L(l_n) = p$ by (2.3).

Consider now the p-periodic sequence (l'_n) defined by $l'_n = g_2(n)$, $n \ge 0$. Note that

$$g_1(n) = g_2(n), \quad 1 \le n \le p - 1,$$

and

$$L_k(l_n) \le L(l'_n) = \frac{p+1}{2}, \quad k \ge 1.$$

Assume now that $1 \le k \le (p-3)/2$. Let (s_n) be any sequence obtained from (l_n) by changing at most $(p-3)/2$ elements. Suppose that g is the polynomial in $\mathbb{F}_p[x]$ of degree at most $p-1$, which represents the sequence (s_n), i.e., $s_n = g(n)$, $n \ge 0$.

It is obvious that the sequences (s_n) and (l'_n) coincide for at least $p - 1 - k \ge (p+1)/2$ elements in a period. Hence, the polynomial $g(X) - g_2(X)$ has at least $(p+1)/2$ zeros, which implies that either $g(X) = g_2(X)$ or $\deg g \ge (p+1)/2$. Therefore $L_k(l_n) = L(l'_n) = (p+1)/2$.

Finally we remark that $L_k(l_n) = 0$ for $k \ge (p-1)/2$, since we have exactly $(p-1)/2$ nonzero elements in a period of (l_n) and the zero sequence of linear complexity 0 can be obtained by $(p-1)/2$ changes. □

Aly and Winterhof also give a lower bound for the k-error linear complexity over $\mathbb{F}_p$ of Sidelnikov sequences in the same paper ,

$$L_k(\sigma_n) \ge \min\left(\left(\frac{p+1}{2}\right)^r - 1, \frac{q-1}{k+1} - \left(\frac{p+1}{2}\right)^r + 1\right).$$

For $k \ge (q-1)/2$ we have $L_k(\sigma_n) = 0$. The 1-error linear complexity over $\mathbb{F}_p$ of Sidelnikov sequences has recently be determined by Eun *et al.* in [20] to be

$$L_1(\sigma_n) = \left(\frac{p+1}{2}\right)^r - 1, \quad q > 3.$$

Other Measures Related to Linear Complexity

The *Kolmogorov complexity* of a binary sequence is, roughly speaking, the length of the shortest computer program that generates the sequence. The relationship between linear complexity and Kolmogorov complexity was studied in [4, 75].

We recall that the *nonlinear complexity profile* $NL_m(s_n, N)$ of an infinite sequence (s_n) over $\mathbb{F}_q$ is the function, which is defined for every integer $N \geq 2$, as the smallest k such that a polynomial recurrence relation

$$s_{n+k} = \Psi(s_{n+k-1}, \ldots, s_n), \qquad 0 \leq n \leq N - k - 1,$$

with a polynomial $\Psi(\lambda_1, \ldots, \lambda_k)$ over $\mathbb{F}_q$ of total degree at most m can generate the first N terms of (s_n). Note that generally speaking $NL_1(s_n, N) \neq L(s_n, N)$ because in the definition of $L(s_n, N)$ one can use only homogeneous linear polynomials. Obviously, we have

$$L(s_n, N) \geq NL_1(s_n, N) \geq NL_2(s_n, N) \geq \ldots.$$

See [32] for the presentation of results on the linear complexity profile of nonlinear, inversive and quadratic exponential generators in a more general form, namely in terms of lower bounds on the nonlinear complexity profile.

Linear Complexity and Predictability

It is clear that sequences with low linear complexity have to be avoided for cryptographic applications. One should note that sequences which show favourable behaviour with respect to linear complexity and related quality measures should also be used with care. Rigorous results, demonstrating this fact, have been recently obtained by Blackburn *et al* [6, 7], which we briefly describe below.

As we have remarked earlier, the inversive generator (2.8) stands out as a sequence with almost best possible linear complexity. Nevertheless it turns out that it is polynomial time predictable if sufficiently many bits of its consecutive terms are known, except for some very limited special cases.

Recall that the inversive generator (y_n) is defined as

$$y_{n+1} = ay_n^{p-2} + b = \begin{cases} ay_n^{-1} + b & \text{if } y_n \neq 0, \\ b & \text{otherwise,} \end{cases} \qquad n \geq 0,$$

with $a, b, y_0 \in \mathbb{F}_p$ (regarded as integers in $\{0, 1, \ldots, p-1\}$), $a \neq 0$.

The elements a, b and y_0 are assumed to be the secret key in the cryptographic setting. Since it is easy to recover the secret key in case several consecutive terms of the sequence are known, it is assumed that only the most significant bits of them are revealed. When approximations x_0, x_1, x_2, x_3 to $y_n, y_{n+1}, y_{n+2}, y_{n+3}$ are known for some n, [6] shows that $a, b, y_n, \ldots, y_{n+3}$ can be recovered in polynomial (in $\log p$) time, if the approximations are sufficiently good and a small set of values of a, b is excluded.

It is shown in [7] that the knowledge of b, and approximations x_0, x_1, x_2 to y_n, y_{n+1}, y_{n+2} is sufficient to recover a and the consecutive terms y_n, y_{n+1}, y_{n+2}, in polynomial time, provided that the approximations are good enough and the initial value y_0 is not in a certain small set. Although the assumption that b is public is not realistic in the cryptographic setting, it is not unlikely that the result can be extended to the case when b is unknown (see [7]).

References to the earlier work on the predictability of linear congruential generators can also be found in [6, 7]. A weaker attack is discussed in Klapper [36], where the idea is to decrease the linear complexity of a given sequence by considering it over a ring which is different from the ring where the sequence is naturally defined (and its high linear complexity is guaranteed). The result in Shparlinski and Winterhof [67] shows that this approach has very limited chance to succeed.

3.   Autocorrelation and Related Distribution Measures for Binary Sequences

3.1   Autocorrelation

One would expect that a periodic random sequence and a shift of it would have a low correlation. Autocorrelation measures the similarity between a sequence (s_n) of period t and its shifts by k positions, for $1 \leq k \leq t - 1$.

The *(periodic) autocorrelation* of a t-periodic binary sequence (s_n) is the function defined by

$$A(s_n, k) = \sum_{n=0}^{t-1} (-1)^{s_{n+k}+s_n}, \quad 1 \leq k \leq t - 1.$$

Note that Section 3 of Chapter 2 is concerned with the *correlation* of two sequences (s_n) and (t_n).

Obviously a low autocorrelation is a desirable feature for pseudorandom sequences that are used in cryptographic systems. Local randomness of periodic sequences is also of importance cryptographically, since only small parts of the period are used for the generation of stream ciphers.

The *aperiodic autocorrelation* reflects local randomness and is defined by

$$AA(s_n, k, u, v) = \sum_{n=u}^{v} (-1)^{s_{n+k}+s_n}, \quad 1 \leq k \leq t - 1, \ 0 \leq u < v \leq p - 1.$$

For the Legendre sequences, for example, $A(l_n, k)$ can be immediately derived from the well-known formula, see e.g. [35],

$$\sum_{n=0}^{p-1} \left(\frac{n}{p}\right) \left(\frac{n+k}{p}\right) = -1, \quad 1 \le k \le p-1,$$

and the following bound on the aperiodic autocorrelation of Legendre sequences follows immediately from (2.11).

Theorem 3.1. *The (aperiodic) autocorrelation of the Legendre sequence satisfies*

$$A(l_n, k) = \left(\frac{k}{p}\right) \left(1 + (-1)^{(p-1)/2}\right) - 1, \quad 1 \le k \le p-1,$$

$$|AA(l_n, k, u, v)| \le 2p^{1/2}(1+\log p) + 2, \ 1 \le k \le p-1, \ 0 \le u \le v \le p-1.$$

For bounds on the aperiodic autocorrelation of extended Legendre sequences see [50]. For the aperiodic autocorrelation of Sidelnikov sequences and two-prime generators see [68] and [10], respectively.

3.2 Related Distribution Measures

Higher Order Correlation

In Mauduit and Sárközy [43] the *correlation measure of order k* of a binary sequence (s_n) is introduced as

$$C_k(s_n) = \max_{M,D} \left| \sum_{n=1}^{M} (-1)^{s_{n+d_1}} \cdots (-1)^{s_{n+d_k}} \right|, \quad k \ge 1,$$

where the maximum is taken over all $D = (d_1, d_2, \ldots, d_k)$ with non-negative integers $d_1 < d_2 < \ldots < d_k$ and M such that $M - 1 + d_k \le T - 1$. $C_2(s_n)$ is obviously bounded by the maximal absolute value of the aperiodic autocorrelation of (s_n). It is also shown in [43] that the Legendre sequence has small correlation measure up to rather high orders.

The following family of pseudorandom binary sequences is introduced in Gyarmati [33]: Let p be an odd prime and g be a primitive root modulo p. Denote by ind n the *discrete logarithm* of n to the base g, i.e., ind $n = j$ if $n = g^j$ with $1 \le j \le p-1$. Let $f(X)$ be a polynomial of degree k modulo p. Then the finite sequence (e_n) is defined by

$$e_n = \begin{cases} 1 & \text{if } 1 \le \text{ind } f(n) \le (p-1)/2, \\ -1 & \text{if } (p+1)/2 \le \text{ind } f(n) \le p-1 \text{ or } p \mid f(n), \end{cases} \quad 1 \le n \le p-1.$$

The correlation measure of the sequence (e_n) is also analysed in [33].

The sequence (k_n) of signs of Kloosterman sums is defined as follows;

$$k_n = \begin{cases} 1 & \text{if } \sum_{j=1}^{p-1} \exp(2\pi i(j + nj^{-1})/p) > 0, \\ -1 & \text{if } \sum_{j=1}^{p-1} \exp(2\pi i(j + nj^{-1})/p) < 0, \end{cases} \quad 1 \le n \le p-1,$$

where j^{-1} is the inverse of j modulo p. Bounds on the correlation measure of order k of (k_n) are given in Fouvry *et al* [22].

Recently Brandstätter and Winterhof [12] have shown that the linear complexity profile of a given t-periodic sequence can be estimated in terms of its correlation measure;

$$L(s_n, N) \ge N - \max_{1 \le k \le L(s_n, N)+1} C_k(s_n), \quad 2 \le N \le t-1.$$

Hence, a lower bound on $L(s_n, N)$ can be obtained whenever an appropriate bound on $\max C_k(s_n)$ is known. The proof is similar to that of Theorem 2.4.

Nonlinearity

Each binary sequence (s_n) of period t over the field $\mathbb{F}_2$ can naturally be associated with a Boolean function B. More precisely, we define an integer m by $2^m \le t < 2^{m+1}$ and denote by $\mathcal{B}_m$ the set of m-bit integers

$$\mathcal{B}_m = \{n \in \mathbb{Z} : 0 \le n \le 2^m - 1\}.$$

We do not distinguish between m-bit integers $n \in \mathcal{B}_m$ and their binary expansion. So $\mathcal{B}_m$ can be considered as the m-dimensional Boolean cube $\mathcal{B}_m = \{0, 1\}^m$. The Boolean function $B : \mathcal{B}_m \to \mathbb{F}_2$ associated to the sequence (s_n) is given by

$$B(n) = s_n, \quad n \in \mathcal{B}_m. \tag{3.1}$$

For $n, r \in \mathcal{B}_m$, $\langle n, r \rangle$ denotes the inner product of n and r considered as binary vectors. That is

$$\langle n, r \rangle = n_1 r_1 + \ldots + n_m r_m,$$

where $n = (n_1, \ldots, n_m)$ and $r = (r_1, \ldots, r_m)$ are the binary representations of n and r.

The *Fourier coefficients* of a Boolean function $B : \mathcal{B}_m \to \{0, 1\}$ are defined as

$$\hat{B}(r) = 2^{-m} \sum_{n \in \mathcal{B}_m} (-1)^{B(n) + \langle n, r \rangle}, \quad r \in \mathcal{B}_m,$$

and the *nonlinearity* $\mathcal{NL}(B)$ is defined as

$$\mathcal{NL}(B) = 2^{m-1} - 2^{m-1} \max_{r \in \mathcal{B}_m} \left| \hat{B}(r) \right|.$$

The nonlinearity corresponds to the smallest possible Hamming distance between the vector of values of B and the vector of values of a linear function in m variables over $\mathbb{F}_2$. For the cryptographic significance of this notion see [11] and references therein. In particular, a high nonlinearity is necessary for achieving confusion and avoiding differential attacks.

In Brandstätter and Winterhof [11] the nonlinearity of the Boolean function B, defined by (3.1) is estimated in terms of the correlation measure of order 2 of the sequence (s_n). It is shown that

$$\mathcal{NL}(B) > 2^{m-1}(1 - 8^{1/4}2^{-m/4}C_2(s_n)^{1/4}).$$

This result can be applied to any binary sequence for which a bound on the correlation measure of order 2 or the aperiodic autocorrelation is known. For example, consider the Boolean function

$$B(n) = \begin{cases} 0 & \text{if } \left(\frac{n}{p}\right) = 1 \text{ or } n = 0, \\ 1 & \text{if } \left(\frac{n}{p}\right) = -1, \end{cases} \quad 0 \le n \le 2^m - 1,$$

corresponding to the Legendre sequence, where $2^m \le p < 2^{m+1}$. The bound

$$\mathcal{NL}(B) = 2^{m-1}(1 + O(2^{-m/8}m^{1/4}))$$

follows immediately from [66, Theorem 10.1] or [43].

Note that the Legendre sequence describes the least significant bit of the discrete logarithm. An analogue result on the nonlinearity of the Boolean function corresponding to the sequence of least significant bits of the discrete logarithm in the finite field $\mathbb{F}_{2^r}$ is given in Brandstätter et al. [9].

4.   Discrepancy and Uniform Distribution

A quantitative measure of uniformity of distribution of a sequence, the so-called discrepancy has a long history. Originated from a classical problem in Diophantine approximations, namely distribution of fractional parts of integer multiples of an irrational in the unit interval, this concept has found applications in various areas like combinatorics, probability theory, mathematical finance, to name a few. It is apparent that it can be used in the analysis of PR sequences; it also emerges as a valuable tool in quasi-Monte Carlo methods where the so-called quasi-random sequences are often utilised.

Let P be a point set (finite sequence) $\mathbf{x}_0, \mathbf{x}_1, \ldots, \mathbf{x}_{N-1} \in [0,1)^s$ with $s \ge 1$. The *discrepancy* $D_N^{(s)}$ of P is defined as

$$D_N^{(s)}(P) = D_N^{(s)}(\mathbf{x}_0, \mathbf{x}_1, \ldots, \mathbf{x}_{N-1}) = \sup_J \left| \frac{A_N(J)}{N} - V(J) \right|,$$

where the supremum is taken over all subboxes $J \subseteq [0, 1)^s$, $A_N(J)$ is the number of points $x_0, x_1, \ldots, x_{N-1}$ in J and $V(J)$ is the volume of J. We put $D_N(P) = D_N^{(1)}(P)$. For an infinite sequence $(s_n) \in [0, 1)^s$, the discrepancy of (s_n) is defined as

$$D_N^{(s)}(s_n) = D_N^{(s)}(s_0, s_1, \ldots, s_{N-1}).$$

It is evident from the well-known Erdős-Turán inequality, (4.1) below, that the main tool in estimating discrepancy is the use of bounds on exponential sums. Let P be a point set $x_0, x_1, \ldots, x_{N-1}$ in $[0, 1)$. There exists an absolute constant C such that for any integer $H \geq 1$,

$$D_N(P) < C \left(\frac{1}{H} + \frac{1}{N} \sum_{h=1}^{H} \frac{1}{h} |S_N(h)| \right), \tag{4.1}$$

where $S_N(h) = \sum_{n=0}^{N-1} \exp(2\pi i h x_n)$.

For the case $s \geq 2$ the generalised version of (4.1), namely the Erdős-Turán-Koksma inequality is used.

The law of the iterated logarithm asserts that the order of magnitude of discrepancy of N points in $[0, 1)^s$ should be around $N^{-1/2}(\log \log N)^{1/2}$. Accordingly, as a measure of randomness of a PRN sequence, one investigates the discrepancy of s-tuples of consecutive terms.

Consider, for example, the inversive congruential PRNs, produced by (2.5), with least period p. For a fixed dimension $s \geq 1$, put

$$x_n = (y_n/p, y_{n+1}/p, \ldots, y_{n+s-1}/p) \in [0, 1)^s, \quad n = 0, \ldots, p-1.$$

Depending on the parameters $a, b \in \mathbb{F}_p$, and in particular on the average, $D_p^{(s)}(x_0, \ldots, x_{p-1})$ has an order of magnitude between $p^{-1/2}$ and $p^{-1/2}(\log p)^s$ for every $s \geq 2$. Similar favourable results are available, for instance, for non-linear, quadratic exponential and elliptic curve generators.

As we have remarked earlier, only parts of the period of a PR sequence are used in applications. Therefore bounds on the discrepancy of sequences in parts of the period are of great interest.

The following theorem of Niederreiter and Shparlinski [57] gives an upper bound for the discrepancy of nonlinear congruential PRNs for parts of the period. We present a slightly imroved version.

Theorem 4.1. *Let (u_n) be a nonlinear congruential generator* (2.7) *with period t. For any positive integer r we have*

$$D_N(u_n/p) = O(N^{-1/(2r)} p^{1/(2r)} (\log p)^{-1/2} \log \log p), \quad 1 \leq N \leq t,$$

where the implied constant depends on r and the degree of f.

Proof. First we prove that, for $\gcd(h, p) = 1$,

$$S_N(h) = O(N^{1-1/(2r)}p^{1/(2r)}(\log p)^{-1/2}), \quad 1 \le N \le t. \quad (4.2)$$

Select any $h \in \mathbb{Z}$ with $\gcd(h, p) = 1$. It is obvious that for any integer $k \ge 0$ we have

$$\left| S_N(h) - \sum_{n=0}^{N-1} e_p(u_{n+k}) \right| \le 2k.$$

Therefore, for any integer $K \ge 1$,

$$K|S_N(H)| \le W + K(K-1),$$

where

$$W = \sum_{n=0}^{N-1} \left| \sum_{k=0}^{K-1} e_p(u_{n+k}) \right|.$$

We consider again the sequence of polynomials $F_k(X)$ defined in the proof of Theorem 2.2. By the Hölder inequality and using $u_{n+k} = F_k(u_n)$ we obtain

$$W^{2r} \le N^{2r-1} \sum_{n=0}^{N-1} \left| \sum_{k=0}^{K-1} e_p(F_k(u_n)) \right|^{2r}$$

$$\le N^{2r-1} \sum_{x \in \mathbb{F}_p} \left| \sum_{k=0}^{K-1} e_p(F_k(x)) \right|^{2r}$$

$$\le N^{2r-1} \sum_{k_1,\ldots,k_{2r}} \left| \sum_{x \in \mathbb{F}_p} e_p(F(x)) \right|,$$

where $F(X) = F_{k_1}(X) + \ldots + F_{k_r}(X) - F_{k_{r+1}}(X) - \ldots - F_{k_{2r}}(X)$. If $\{k_1, \ldots, k_r\} = \{k_{r+1}, \ldots, k_{2r}\}$, then $F(X)$ is constant and the inner sum is trivially equal to p. There are at most $r!K^r$ such sums. Otherwise we can apply Weil's bound to the inner sum using $\deg F \le d^{K-1}$, to get the upper bound $d^{K-1}p^{1/2}$ for at most K^{2r} sums. Hence,

$$W^{2r} \le r!K^r N^{2r-1}p + d^{K-1}K^{2r}N^{2r-1}p^{1/2}.$$

Choose

$$K = \left\lceil 0.4 \frac{\log p}{\log d} \right\rceil.$$

Then it is easy to see that the first term dominates the second one and we get (4.2) after simple calculations. Choosing

$$H = \left\lceil N^{1/(2r)}p^{-1/(2r)}(\log p)^{1/2} \right\rceil$$

in (4.1), we obtain the discrepancy bound.   □

Note that the known upper bounds obtained for full period are often the best possible, as the corresponding lower bounds demonstrate (see [53]). However for parts of the period, the bound in Theorem 4.1 is rather weak and improvements are sought for. One should note on the other hand that the method used in [57] for estimating $S_N(h)$, is the first to give nontrivial bounds for parts of the period. This method also applies in case $s \geq 2$.

As to bounds on discrepancy of some special nonlinear generators for parts of the period, much better results can be obtained. For the inversive congruential generators (y_n) of period t, Niederreiter and Shparlinski showed in [58] that

$$D_N(y_n/p) = O(N^{-1/2}p^{1/4}\log p), \quad 1 \leq N \leq t.$$

For an average discrepancy bound over all initial values of a fixed inversive congruential generator see Niederreiter and Shparlinski [60].

Results about the distribution of the power generator follow from the bounds of exponential sums in Friedlander and Shparlinski [25] and in Bourgain [8]. Exponential sums of nonlinear generators with Dickson polynomials have been estimated in Gomez-Perez *et al* [28]. Discrepancy bounds for nonlinear congruential generators of higher order can be found in [29, 31, 72].

For the distribution of explicit nonlinear generators see the series of papers [61, 62, 77]. In particular for the explicit inversive generator (2.4) we have the discrepancy bound

$$D_N(z_n/p) = O(\min\{N^{-1/2}p^{1/4}\log p, N^{-1}p^{1/2}(\log p)^2\}), \quad 1 \leq N \leq p.$$

The order of magnitude of discrepancy of the PRNs produced by the elliptic curve generator of period t with $f(x,y) = x$ or $f(x,y) = y$ is $t^{-1}p^{1/2}\log p$, by Hess and Shparlinski [34]. This result can be easily extended to parts of the period. We present the proof of the following special version.

Theorem 4.2. *The sequence* (w_n) *defined by* (2.15), *having period t satisfies*

$$D_N(w_n/p) = O(t^{-1}p^{1/2}\log p\log t), \quad 1 \leq N < t.$$

Proof. First we estimate the exponential sums

$$S_N = \sum_{n=1}^{N-1} e_p(w_n) = \sum_{n=1}^{N-1} e_p(x(nG)), \quad 1 \leq N < t.$$

Using the Vinogradov method again we get by (2.12)

$$|S_N| \leq \frac{1}{t}\sum_{a=0}^{t-1}\left|\sum_{n=1}^{t-1} e_p(x(nG))e_t(an)\right|\left|\sum_{m=0}^{N-1} e_t(am)\right| = O(p^{1/2}\log t)$$

by (2.13) and [38, Corollary 1]. The discrepancy bound follows from (4.1). □

The distribution of an elliptic curve analogue of the power generator has been analysed in Lange and Shparlinski [40].

We should remark that the linear congruential generator, unlike other generators we mentioned above, is distributed too evenly. In case a in (2.1) is a primitive root mod p, $b = 0$ and $s_0 \neq 0$, the sequence has period length $p - 1$, and for most choices of a,

$$D^{(s)}_{p-1}(s_n/p) = O(p^{-1}(\log p)^s(\log\log(p+1))).$$

Although such low-discrepancy sequences need to be avoided as PR sequences, they are needed for use in quasi-Monte Carlo methods (see [53]). The study of *irregularities of distribution* suggests that for any N-element point set P and any sequence (s_n) in $[0, 1)^s$, $s \geq 1$, the least order of magnitude of $D^{(s)}_N(P)$ and $D^{(s)}_N(s_n)$ can be $N^{-1}(\log n)^{s-1}$ and $N^{-1}(\log n)^s$, respectively.

The construction of point sets and sequences with these least possible bounds has been a challenging problem; both for theoretical interest and for applications. Recent results of Xing, Niederreiter and Özbudak show that geometric methods are particularly fruitful for such constructions. We refer the reader to the surveys by Niederreiter [54, 56] for an extensive description of this study, illustrating yet another application of global function fields.

References

[1] H. Aly and A. Winterhof, "On the linear complexity profile of nonlinear congruential pseudorandom number generators with Dickson polynomials", Des. Codes Cryptogr., to appear.

[2] H. Aly and A. Winterhof, "On the k-error linear complexity over $\mathbb{F}_p$ of Legendre and Sidelnikov sequences", preprint 2005.

[3] P. Beelen and J.M. Doumen, "Pseudorandom sequences from elliptic curves", *Finite fields with applications to coding theory, cryptography and related areas (Oaxaca, 2001)*, Springer, Berlin, 37–52 (2002).

[4] T. Beth and Z.D. Dai, "On the complexity of pseudo-random sequences—or: If you can describe a sequence it can't be random", *Advances in cryptology—EUROCRYPT '89 (Houthalen, 1989)*, Lecture Notes in Comput. Sci., Vol. 434, 533–543 (1990).

[5] S.R. Blackburn, T. Etzion and K.G. Paterson, "Permutation polynomials, de Bruijn sequences, and linear complexity", J. Combin. Theory Ser. A, Vol. 76, 55–82 (1996).

[6] S.R. Blackburn, D. Gomez-Perez, J. Gutierrez and I. Shparlinski, "Predicting the inversive generator", Lecture Notes in Comput. Sci., Vol. 2898, 264–275 (2003).

[7] S.R. Blackburn, D. Gomez-Perez, J. Gutierrez and I. Shparlinski, "Predicting nonlinear pseudorandom number generators", Math. Comp., Vol. 74, 1471–1494 (2005).

[8] J. Bourgain, "Mordell's exponential sum estimate revisited", J. Amer. Math. Soc., Vol. 18, 477-499 (2005).

[9] N. Brandstätter, T. Lange and A. Winterhof, "On the non-linearity and sparsity of Boolean functions related to the discrete logarithm", preprint 2005.

[10] N. Brandstätter and A. Winterhof, "Some notes on the two-prime generator", IEEE Trans. Inform. Theory, Vol. 51, 3654–3657 (2005).

[11] N. Brandstätter and A. Winterhof, "Nonlinearity of binary sequences with small autocorrelation", *Proceedings of the Second International Workshop on Sequence Design and its Applications in Communications (IWSDA'05)*, to appear.

[12] N. Brandstätter and A. Winterhof, "Linear complexity profile of binary sequences with small correlation measure", preprint 2005.

[13] T.W. Cusick, C. Ding and A. Renvall, *Stream ciphers and number theory*, Revised edition. North-Holland Mathematical Library, 66. Elsevier Science B.V., Amsterdam, 2004.

[14] C. Ding, G. Xiao and W. Shan, *The stability theory of stream ciphers*, Lecture Notes in Computer Science, Vol. 561, Springer-Verlag, Berlin 1991.

[15] G. Dorfer, W. Meidl and A. Winterhof, "Counting functions and expected values for the lattice profile at n", Finite Fields Appl., Vol. 10, 636–652 (2004).

[16] G. Dorfer and A. Winterhof, "Lattice structure and linear complexity profile of nonlinear pseudorandom number generators", Appl. Algebra Engrg. Comm. Comput., Vol. 13, 499–508 (2003).

[17] J. Eichenauer, H. Grothe, J. Lehn and A. Topuzoğlu, "A multiple recursive nonlinear congruential pseudo random number generator", Manuscripta Math., Vol. 59, 331–346 (1987).

[18] J. Eichenauer and J. Lehn, "A nonlinear congruential pseudorandom number generator", Statist. Hefte, Vol. 27, 315–326 (1986).

[19] J. Eichenauer-Herrmann, "Statistical independence of a new class of inversive congruential pseudorandom numbers", Math. Comp., Vol. 60, 375–384 (1993).

[20] Y.-C. Eun, H.-Y. Song and M.G. Kyureghyan, "One-error linear complexity over $\mathbb{F}_p$ of Sidelnikov sequences", *Sequences and Their Applications SETA 2004*, Lecture Notes in Comput. Sci., Vol. 3486, 154–165 (2005).

[21] M. Flahive and H. Niederreiter, "On inversive congruential generators for pseudorandom numbers", *Finite fields, coding theory, and advances in communications and computing (Las Vegas, NV, 1991)*, Lecture Notes in Pure and Appl. Math., Vol. 141, 75–80 (1993).

[22] É. Fouvry, P. Michel, J. Rivat and A. Sárközy, "On the pseudorandomness of the signs of Kloosterman sums", J. Aust. Math. Soc., Vol. 77, 425–436 (2004).

[23] J.B. Friedlander, C. Pomerance and I. Shparlinski, "Period of the power generator and small values of Carmichael's function", Math. Comp., Vol. 70, 1591–1605 (2001).

[24] J.B. Friedlander, C. Pomerance and I. Shparlinski, "Corrigendum to: Period of the power generator and small values of Carmichael's function", Math. Comp., Vol. 71, 1803–1806 (2002).

[25] J.B. Friedlander and I. Shparlinski, "On the distribution of the power generator", Math. Comp., Vol. 70, 1575–1589 (2001).

[26] R.A. Games and A.H. Chan, "A fast algorithm for determining the complexity of a binary sequence with period 2^n", IEEE Trans. Inform. Theory, Vol. 29, 144–146 (1983).

[27] M.Z. Garaev, F. Luca, I. Shparlinski and A. Winterhof, "On the lower bound of the linear complexity over $\mathbb{F}_p$ of Sidelnikov sequences", preprint 2005.

[28] D. Gomez-Perez, J. Gutierrez and I. Shparlinski, "Exponential sums with Dickson polynomials", Finite Fields Appl., Vol. 12, 16-25 (2006).

[29] F. Griffin, H. Niederreiter and I. Shparlinski, "On the distribution of nonlinear recursive congruential pseudorandom numbers of higher orders", *Applied algebra, algebraic algorithms and error-correcting codes (Honolulu, HI, 1999)*, Lecture Notes in Comput. Sci., Vol. 1719, 87–93 (1999).

[30] F. Griffin and I. Shparlinski,"On the linear complexity profile of the power generator", IEEE Trans. Inform. Theory, Vol. 46, 2159–2162 (2000).

[31] J. Gutierrez and D. Gomez-Perez, "Iterations of multivariate polynomials and discrepancy of pseudorandom numbers", *Applied algebra, algebraic algorithms and error-correcting codes (Melbourne, 2001)*, Lecture Notes in Comput. Sci., Vol. 2227, 192–199 (2001).

[32] J. Gutierrez, I. Shparlinski and A. Winterhof, "On the linear and nonlinear complexity profile of nonlinear pseudorandom number-generators", IEEE Trans. Inform. Theory, Vol. 49, 60–64 (2003).

[33] K. Gyarmati, "On a family of pseudorandom binary sequences", Period. Math. Hungar., Vol. 49, 45–63 (2004).

[34] F. Hess and I. Shparlinski, "On the linear complexity and multidimensional distribution of congruential generators over elliptic curves", Des. Codes and Cryptogr., Vol. 35, 111–117 (2005).

[35] D. Jungnickel, *Finite fields. Structure and arithmetics*, Bibliographisches Institut, Mannheim, 1993.

[36] A. Klapper, "The vulnerability of geometric sequences based on fields of odd characteristic", J. Cryptology, Vol. 7, 33–51 (1994).

[37] N. Koblitz, *Algebraic Aspects of Cryptography*, Springer-Verlag, Berlin Heidelberg, 1998.

[38] D.R. Kohel and I. Shparlinski, "On exponential sums and group generators for elliptic curves over finite fields", *Algorithmic number theory (Leiden, 2000)*, Lecture Notes in Comput. Sci., Vol. 1838, 395–404 (2000).

[39] S. Konyagin, T. Lange and I. Shparlinski, "Linear complexity of the discrete logarithm", Des. Codes Cryptogr., Vol. 28, 135–146 (2003).

[40] T. Lange and I. Shparlinski, "Certain exponential sums and random walks on elliptic curves", Canad. J. Math., Vol. 57, 338–350 (2005).

[41] R. Lidl, G.L. Mullen and G. Turnwald, *Dickson polynomials*, Pitman Monographs and Surveys in Pure and Applied Mathematics, 65. Longman Scientific & Technical, Harlow; copublished in the United States with John Wiley & Sons, Inc., New York, 1993.

[42] J.L. Massey and S. Serconek, "Linear complexity of periodic sequences: a general theory", *Advances in cryptology—CRYPTO '96 (Santa Barbara, CA)*, Lecture Notes in Comput. Sci., Vol. 1109, 358–371 1996.

[43] C. Mauduit and A. Sárközy, "On finite pseudorandom binary sequences. I. Measure of pseudorandomness, the Legendre symbol", Acta Arith., Vol. 82, 365–377 (1997).

[44] W. Meidl, "How many bits have to be changed to decrease the linear complexity?", Des. Codes Cryptogr., Vol. 33, 109–122 (2004).

[45] W. Meidl, "Linear complexity and k-error linear complexity for p^n-periodic sequences", *Coding, cryptography and combinatorics*, Progr. Comput. Sci. Appl. Logic, Vol. 23, 227–235 (2004).

[46] W. Meidl and A. Winterhof, "Lower bounds on the linear complexity of the discrete logarithm in finite fields", IEEE Trans. Inform. Theory, Vol. 47, 2807–2811 (2001).

[47] W. Meidl and A. Winterhof, "Linear complexity and polynomial degree of a function over a finite field", *Finite fields with applications to coding theory, cryptography and related areas (Oaxaca, 2001)*, Springer, Berlin, 229–238 (2002).

[48] W. Meidl and A. Winterhof, "On the linear complexity profile of explicit nonlinear pseudorandom numbers", Inform. Process. Lett., Vol. 85, 13–18 (2003).

[49] W. Meidl and A. Winterhof, "On the linear complexity profile of some new explicit inversive pseudorandom numbers", J. Complexity, Vol. 20, 350–355 (2004).

[50] W. Meidl and A. Winterhof, "On the autocorrelation of cyclotomic generators", *Finite fields and applications*, Lecture Notes in Comput. Sci., Vol. 2948, 1–11 (2004).

[51] W. Meidl and A. Winterhof, "Some notes on the linear complexity of Sidelnikov-Lempel-Cohn-Eastman sequences", Designs, Codes and Cryptography, to appear.

[52] W. Meidl and A. Winterhof, "On the linear complexity profile of nonlinear pseudorandom number generators with Rédei functions", Finite Fields Appl., to appear.

[53] H. Niederreiter, *Random number generation and quasi-Monte Carlo methods*, CBMS-NSF Regional Conference Series in Applied Mathematics, Vol. 63. Society for Industrial and Applied Mathematics (SIAM), Philadelphia, PA, 1992.

[54] H. Niederreiter, "Constructions of (t, m, s)-nets", *Monte Carlo and quasi-Monte Carlo methods 1998 (Claremont, CA)*, Springer, Berlin, 70–85 (2000).

[55] H. Niederreiter, "Linear complexity and related complexity measures for sequences", *Progress in cryptology—INDOCRYPT 2003*, Lecture Notes in Comput. Sci., Vol. 2904, 1–17 (2003).

[56] H. Niederreiter, "Constructions of (t, m, s)-nets and (t, s)-sequences", Finite Fields Appl., Vol. 11, 578–600 (2005).

[57] H. Niederreiter and I. Shparlinski, "On the distribution and lattice structure of nonlinear congruential pseudorandom numbers", Finite Fields Appl., Vol. 5, 246–253 (1999).

[58] H. Niederreiter and I. Shparlinski, "On the distribution of inversive congruential pseudorandom numbers in parts of the period", Math. Comp., Vol. 70, 1569–1574 (2001).

[59] H. Niederreiter and I. Shparlinski, "Recent advances in the theory of nonlinear pseudorandom number generators", *Monte Carlo and quasi-Monte Carlo methods, 2000 (Hong Kong)*, Springer, Berlin, 86–102 (2002).

[60] H. Niederreiter and I. Shparlinski, "On the average distribution of inversive pseudorandom numbers", Finite Fields Appl., Vol. 8, 491–503 (2002).

[61] H. Niederreiter and A. Winterhof, "Incomplete exponential sums over finite fields and their applications to new inversive pseudorandom number generators", Acta Arith., Vol. 93, 387–399 (2000).

[62] H. Niederreiter and A. Winterhof, "On the distribution of some new explicit nonlinear congruential pseudorandom numbers", *Sequences and Their Applications SETA 2004*, Lecture Notes in Comput. Sci., Vol. 3486, 266–274, (2005).

[63] R. Nöbauer, "Rédei-Permutationen endlicher Körper", *Contributions to general algebra, 5 (Salzburg, 1986)*, Hölder-Pichler-Tempsky, Vienna, 235–246 (1987).

[64] W.M. Schmidt, *Equations over finite fields. An elementary approach*, Lecture Notes in Mathematics, Vol. 536, 1976.

[65] I. Shparlinski, "On the linear complexity of the power generator", Des. Codes Cryptogr., Vol. 23, 5–10 (2001).

[66] I. Shparlinski, *Cryptographic applications of analytic number theory. Complexity lower bounds and pseudorandomness*, Progress in Computer Science and Applied Logic, 22. Birkhäuser Verlag, Basel, 2003.

[67] I. Shparlinski and A. Winterhof, "On the linear complexity of bounded integer sequences over different moduli", Inform. Process. Lett., Vol. 96, 175-177 (2005).

[68] V.M. Sidel'nikov, "Some k-valued pseudo-random sequences and nearly equidistant codes", Problems of Information Transmission, Vol. 5, 12–16 (1969); translated from Problemy Peredači Informacii, Vol. 5, 16–22 (1969), (Russian).

[69] M. Stamp and C.F. Martin, "An algorithm for the k-error linear complexity of binary sequences with period 2^n", IEEE Trans. Inform. Theory, Vol. 39, 1398–1401 (1993).

[70] A. Tietäväinen, "Vinogradov's method and some applications", *Number theory and its applications (Ankara, 1996)*, Lecture Notes in Pure and Appl. Math., Vol. 204, 261–282 (1999).

[71] A. Topuzoğlu and A. Winterhof, "On the linear complexity profile of nonlinear congruential pseudorandom number generators of higher orders", Appl. Algebra Engrg. Comm. Comput., Vol. 16, 219–228 (2005).

[72] A. Topuzoğlu and A. Winterhof, "On the distribution of inversive congruential pseudorandom number generators of higher orders with largest possible period", preprint 2006.

[73] R.J. Turyn, "The linear generation of Legendre sequence", J. Soc. Indust. Appl. Math., Vol. 12, 115–116 (1964).

[74] I.M. Vinogradov, *Elements of number theory*, Dover Publications, Inc., New York, 1954.

[75] Y. Wang, "Linear complexity versus pseudorandomness: on Beth and Dai's result", *Advances in cryptology—ASIACRYPT'99 (Singapore)*, Lecture Notes in Comput. Sci., Vol. 1716, 288–298 (1999).

[76] A. Winterhof, "A note on the linear complexity profile of the discrete logarithm in finite fields", *Coding, cryptography and combinatorics*, Progr. Comput. Sci. Appl. Logic, Vol. 23, 359–367 (2004).

[77] A. Winterhof, "On the distribution of some new explicit inversive pseudorandom numbers and vectors", *Proceedings MC2QMC 2004*, to appear.

[78] G. Xiao and S. Wei, "Fast algorithms for determining the linear complexity of period sequences", *Progress in cryptology - INDOCRYPT 2002. Third international conference on cryptology in India, Hyderabad, India, December 16-18, 2002*, Lect. Notes in Comput. Sci., Vol. 2551, 12-21, (2002).

[79] G. Xiao, S. Wei, K.Y. Lam and K. Imamura, "A fast algorithm for determining the linear complexity of a sequence with period p^n over $GF(q)$", IEEE Trans. Inform. Theory, Vol. 46, 2203–2206 (2000).

Chapter 5

GROUP STRUCTURE OF ELLIPTIC CURVES OVER FINITE FIELDS AND APPLICATIONS

Ram Murty and Igor Shparlinski

1. Introduction

1.1 Motivation and General Outline

Let $\mathbf{E}$ be an elliptic curve defined over $\mathbb{F}_q$, the finite field of q elements. It is known that the set of $\mathbb{F}_q$-rational points of $\mathbf{E}$ has a structure of an abelian group. This fact, since the works of Koblitz [68] and Miller [98], underlies all known applications of elliptic curves to cryptography, see [3, 15, 16, 50, 73] and references therein. We give a survey of recent results about the structure of this group as well as techniques used.

In particular, we discuss how often this group is cyclic or has a massive cyclic component. The famous *Lang–Trotter conjecture* [77], which in turn is motivated by *Artin's conjecture*, is a typical example of what kind of question we consider. We recall, that Artin's conjecture, in a quantitative form, claims that for any integer $g \neq -1$, which is not a perfect square, there is a constant $A(g) > 0$ such that g is a primitive root for $N_g(x) \sim A(g)\pi(x)$ primes $p \leq x$, where as usual, $\pi(x)$ is the total number of primes $p \leq x$. Under the *Extended Riemann Hypothesis* (ERH), this conjecture has been proved by Hooley [61]. There are also several unconditional results on Artin's conjecture "on average" over g, see Chapter 2 in [110] and a more recent survey [84]. There is also a series of results, originated by the work of Murty [103], and then further developed in [55, 59, 109], which show that the multiplicative structure of possible exceptional values of g is very restrictive.

Nowadays a great variety of modifications, generalizations and applications of Artin's conjecture is known, see [65, 85, 104, 112], among which one can

A. Garcia and H. Stichtenoth (eds.), Topics in Geometry, Coding Theory and Cryptography, 167–198.
© 2007 *Springer.*

consider elliptic curve analogues of Artin's conjecture, see [105]. In particular, one can ask for how many primes $p \leq x$, the reduction modulo p of a fixed point G of infinite order on a given elliptic curve $\mathbf{E}$ defined over $\mathbb{Q}$ generates the whole group of points on the reduction of $\mathbf{E}$ modulo p. The Lang–Trotter conjecture predicts an asymptotic formula for this quantity. A related question is the question of Serre [123] about counting for how many primes $p \leq x$, this reduction of $\mathbf{E}$ modulo p forms a cyclic group. We discuss these questions in Section 2.1.

An associated question is the question about the largest cyclic component of the group of points on an elliptic curve over a finite field. This and some other related questions are addressed in Section 2.2.

We also describe several recent results about the arithmetic structure of the cardinality of this group (which is ultimately related to the group structure), see Section 2.3.

These questions are of intrinsic interest and also have very important applications to cryptography which we outline too. We do not discuss the discrete logarithm problem on elliptic curves and various attacks on this problem but rather refer to [3, 15, 16, 50, 73] where the reader can find exhaustive treatments of this topic. Instead, we concentrate on less known questions about relevance of the results on the group structure of the set of $\mathbb{F}_q$-rational points on an elliptic curve and the arithmetic properties of the cardinality of this set for several recently emerged applications to cryptography and pose some open questions for further research. For instance, we describe some recent results characterising the probability of success of the so-called **MOV** attack on the discrete logarithm problem on elliptic curves, see Section 3.1. Very often, such questions are extremely hard. For example, the question about the primality of the number of $\mathbb{F}_q$-rational points on an elliptic curve is of great cryptographic significance but appears to be out of reach at the present time.

The wealth of various techniques which have been used to treat these questions is truly remarkable. They use tools from essentially all areas of number theory and include, but are not limited to:

- ERH

- Chebotarev Density Theorem

- Bounds of character and exponential sums

- L-functions

- Subspace Theorem

- Linear forms in p-adic logarithms

- Distribution of prime numbers

and many others.

Unfortunately, we are not able to provide complete or even abbreviated proofs of the results we discuss, which are typically quite technically involved and lengthy (and as we have mentioned require rather extensive background from other areas of number theory). However sometimes we try to explain the main ideas behind these proofs and refer the reader to the original publications for further details.

Neither do we give a systematic introduction to elliptic curves, but rather refer to some standard sources such as [127]. We however give a self-contained outline of the basic facts on elliptic curves which are used here.

We would like to use this survey as an opportunity to attract more attention of the cryptographic community to a great variety of readily available number theoretic results and techniques which can be of great significance for cryptography. On the other hand, we hope that more number theorists can be motivated to use their skills for solving problems of cryptographic significance. There are many problems which are important for cryptography and for which heuristic approaches are not immediately clear and thus rigorous analysis is necessary and appears to be rather feasible.

Throughout the paper, any implied constants in the symbols "O", " $\ll$ " and " $\gg$ " may occasionally depend (where obvious) on the elliptic curve $\mathbf{E}$ when it is fixed and some other parameters vary and are absolute otherwise. We recall that the statements $A \ll B$ and $B \gg A$ are equivalent to $A = O(B)$ for positive functions A and B.

1.2 Background on Elliptic Curves

Let $\mathbf{E}$ be an elliptic curve defined over $\mathbb{F}_q$, defined by an *affine Weierstrass equation*

$$Y^2 + a_1 XY + a_3 Y = X^3 + a_2 X^2 + a_4 X + a_6$$

with the *discriminant* $D_{\mathbf{E}} \neq 0$ (an explicit, but rather lengthy, expression for $D_{\mathbf{E}}$ can be found in many standard texts on elliptic curves, for example, see [127]).

Sometimes it is more convenient to consider the projective model, but here we prefer the affine representation.

Two elliptic curves $\mathbf{E}_1$ and $\mathbf{E}_2$ defined over $\mathbb{F}_q$ are called *isomorphic* if there exist two morphisms (that is, two rational maps, regular at any point) $\varphi : \mathbf{E}_1 \to \mathbf{E}_2$ and $\psi : \mathbf{E}_2 \to \mathbf{E}_1$ such that their compositions $\psi\varphi$ and $\varphi\psi$ are the identity maps on $\mathbf{E}_1$ and $\mathbf{E}_2$, respectively.

Let $N_{\mathbf{E}}(q)$ denote the number of $\mathbb{F}_q$-rational points on an elliptic curve $\mathbf{E}$ defined over $\mathbb{F}_q$, including one more special point which is called the *point at infinity* $\mathcal{O}$ (which cannot be described in affine coordinates, but has a projective representation).

Two elliptic curves $\mathbf{E}_1$ and $\mathbf{E}_2$ defined over $\mathbb{F}_q$ are called *isogenous* if $N_{\mathbf{E}_1}(q) = N_{\mathbf{E}_2}(q)$.

If the characteristic of $\mathbb{F}_q$ is $p > 3$ then each curve $\mathbf{E}$ has an isomorphic version which can be described by a Weierstrass equation of the form

$$Y^2 = X^3 + AX + B. \tag{1.1}$$

In this case $D_{\mathbf{E}} = -16(4A^3 + 27B^2)$.

In particular we see that the set of all possible elliptic curves $\mathcal{E}(q)$ defined by all possible equations (1.1) over $\mathbb{F}_q$ contains

$$\#\mathcal{E}(q) = q^2 + O(q)$$

curves (note that here we count the number of distinct Weierstrass equations rather than the number of isomorphism classes of elliptic curves). However, each curve has approximately the same number of isomorphic copies in this set defined by the *twists* of the original equation

$$Y^2 = X^3 + At^4 X + Bt^6,$$

(the number of isomorphic twists depends on $\gcd(q-1, 12)$ and whether $AB = 0$ or not).

We also consider extensions $\mathbb{F}_{q^n}$ of $\mathbb{F}_q$ and, accordingly, we consider the sets $\mathbf{E}(\mathbb{F}_{q^n})$ of $\mathbb{F}_{q^n}$-rational points on $\mathbf{E}$ (including the point at infinity $\mathcal{O}$).

On the other hand, given an elliptic curve $\mathbf{E}$ defined over $\mathbb{Q}$ we consider sets of $\mathbb{F}_p$-rational points on the reduction of this curve modulo p.

We recall that $\mathbf{E}(\mathbb{F}_{q^n})$ forms an abelian group (with $\mathcal{O}$ as the identity element). The cardinality $N_{\mathbf{E}}(q^n) = \#\mathbf{E}(\mathbb{F}_{q^n})$ of this group satisfies the Hasse–Weil inequality

$$|N_{\mathbf{E}}(q^n) - q^n - 1| \leq 2q^{n/2} \tag{1.2}$$

(see [15, 127, 128] for this, and other general properties of elliptic curves).

We remark that two isomorphic curves are also isomorphic as abelian groups, but the converse statement is not true.

It is well-known that the group of $\mathbb{F}_{q^n}$-rational points $\mathbf{E}(\mathbb{F}_{q^n})$ is of the form

$$\mathbf{E}(\mathbb{F}_{q^n}) \cong \mathbb{Z}/L\mathbb{Z} \times \mathbb{Z}/M\mathbb{Z}, \tag{1.3}$$

where the integers L and M are uniquely determined with $M \mid L$. In particular, $N_{\mathbf{E}}(q^n) = LM$. The number $\ell_{\mathbf{E}}(q^n) = L$ is called the *exponent* of $\mathbf{E}(\mathbb{F}_{q^n})$, and is the largest possible order of points $P \in \mathbf{E}(\mathbb{F}_{q^n})$.

In this paper we give a survey of recent results on the group structure of $\mathbf{E}(\mathbb{F}_{q^n})$.

Trivially, from the definition (1.3), and from (1.2), we see that the inequality

$$\ell_{\mathbf{E}}(q^n) \geq (N_{\mathbf{E}}(q^n))^{1/2} \geq (q^n + 1 - 2q^{n/2})^{1/2} = q^{n/2} - 1$$

holds for all q and n. It is not hard to show that this bound can be attained.

The difference

$$a_{\mathbf{E}}(q^n) = N_{\mathbf{E}}(q^n) - q^n - 1$$

plays a prominent role in determining the properties of $\mathbf{E}$. In particular, we recall that an elliptic curve $\mathbf{E} \in \mathcal{E}(q)$ is called *supersingular* if and only if $a_{\mathbf{E}}(q) \equiv 0 \pmod{p}$ (this is one of many alternative definitions, see [15, 127]). Otherwise, $\mathbf{E}$ is called *non-supersingular* or *ordinary*.

If $\mathbf{E}$ is an elliptic curve over an algebraic number field $\mathbb{K}$, the endomorphism ring of $\mathbf{E}$ over $\mathbb{K}$, denoted $\mathrm{End}_{\mathbb{K}}(\mathbf{E})$, contains a copy of the integers corresponding to the morphisms $x \mapsto nx$ for each $n \in \mathbb{Z}$. If this ring is strictly bigger than $\mathbb{Z}$, we say $\mathbf{E}$ has *complex multiplication* (CM) for in that case, it is a classical theorem that the ring is isomorphic to an order in an imaginary quadratic field. Otherwise, we say $\mathbf{E}$ is a non-CM curve. In the case $\mathbb{K}$ is a finite field, the ring of endomorphisms is always larger than $\mathbb{Z}$ and two cases can arise. Either the ring is isomorphic to an order in an imaginary quadratic field or it is an order in a quaternion algebra. In the latter case, this is equivalent to $\mathbf{E}$ having supersingular reduction. Many of the questions about elliptic curves fall naturally into these two categories, the CM case and the non-CM case. In many instances, the CM case is the easier one to deal with since we can then exploit the presence of additional structure.

Although generally the group structure of elliptic curves over finite fields, described by (1.3) is reasonably well understood, still there are many delicate aspects and challenging open questions. For instance, we describe several known results about the size and arithmetic structure of the integers M and L which appear in (1.3). Here, we are mainly interested in a better lower bound of the size of the exponent $\ell_{\mathbf{E}}(q^n)$ when at least one of the parameters $\mathbf{E}$, q and n varies. In particular, the case of cyclicity $\ell_{\mathbf{E}}(q^n) = N_{\mathbf{E}}(q^n)$ is of primary interest.

2.   Group Structure

2.1   Cyclicity

Here we concentrate on the question of cyclicity, that is, whether $\ell_{\mathbf{E}}(q^n) = N_{\mathbf{E}}(q^n)$, which essentially dates back to work of Borosh, Moreno and Porta [20] as well as Serre [123].

In the situation where $\mathbf{E}$ is defined over $\mathbb{Q}$, the question about the cyclicity of the reduction $\mathbf{E}(\mathbb{F}_p)$ when p runs over the primes has been addressed in a series of recent works [24, 25, 29, 30] due to Cojocaru, Fouvry and Murty, see also a recent survey [26]. In particular, this problem is closely related to the famous *Lang–Trotter conjecture*, see [77].

One can show that $\mathbf{E}(\mathbb{F}_p)$ is cyclic if and only if p does not split completely in each of the fields $\mathbb{K}_\ell = \mathbb{Q}(\mathbf{E}[\ell])$ obtained from $\mathbb{Q}$ by adjoining to it the co-ordinates of the ℓ-division points of $\mathbf{E}$, as ℓ varies over the primes. Thus, an application of the inclusion-exclusion principle along with the effective *Chebotarev Density Theorem* with the error term implied by the ERH allows one to easily derive an asymptotic formula for the number of such primes (see [103]). Structurally, this problem is similar to Artin's primitive root conjecture in that the same method carries over. However, there are some major differences as has been noted in [57]. The most notable is the observation that $\mathbb{K}_\ell$ contains the ℓ-th cyclotomic field generated over $\mathbb{Q}$ by a primitive ℓ-th root of unity $\zeta_\ell = \exp(2\pi i/\ell)$. Thus, if p splits completely in $\mathbb{K}_\ell$, then it splits completely in $\mathbb{Q}(\zeta_\ell)$. Primes that split completely in the latter field are necessarily congruent to 1 modulo ℓ. This implies that if we can restrict the number of prime divisors of $p - 1$ then there are fewer cases to check in the inclusion-exclusion procedure. Restricting the number of prime divisors of $p - 1$ can be achieved by an application of the lower bound sieve and this strategy has been successfully carried out by Gupta and Murty in [57], where they have shown that if $\mathbf{E}$ has an irrational 2-division point, then there are at least

$$\pi_\mathbf{E}(X) \gg \frac{x}{(\log x)^2}$$

primes $p \leq x$ for which $\mathbf{E}(\mathbb{F}_p)$ is cyclic, where the implied constant depends only on the curve $\mathbf{E}$.

Another feature different from the setting of the Artin primitive root question is highlighted in the paper of Cojocaru and Murty [30] where they prove, assuming the ERH, that

$$\pi_\mathbf{E}(X) = c_\mathbf{E} \operatorname{li} x + O(x^{5/6} \log^{2/3} x)$$

with $c_\mathbf{E} > 0$ whenever $\mathbf{E}$ has an irrational 2-division point. It is unclear if such error terms can be derived in the Artin primitive root conjecture setting assuming only the ERH.

The elliptic curve analog of the primitive root conjecture predicts in the generic case an infinitude of primes for which $\mathbf{E}(\mathbb{F}_p)$ is generated by the modulo p reduction of a fixed global rational point. This is still unresolved even under the assumption of the ERH. However, in the CM case, Gupta and Murty [56] have been able to prove the conjecture under the ERH by exploiting the fact that $\mathbb{K}_\ell$ is abelian over the CM field of $\mathbf{E}$ in this situation.

For curves in extension fields, this question appears to be somewhat easier and has been satisfactorily answered by Vlăduţ [134]. For example given a curve $\mathbf{E}$ over $\mathbb{F}_q$, and a large real $x > 0$ one can define the set

$$\mathcal{C}_\mathbf{E}(x) = \{n \leq x \mid \mathbf{E}(\mathbb{F}_{q^n}) \text{ is cyclic}\}.$$

Then we can define the upper and lower densities

$$D(\mathbf{E}) = \limsup_{x \to \infty} \#\mathcal{C}_\mathbf{E}(x)/x \qquad \text{and} \qquad d(\mathbf{E}) = \liminf_{x \to \infty} \#\mathcal{C}_\mathbf{E}(x)/x$$

Clearly $D(\mathbf{E}) = d(\mathbf{E}) = 0$ unless $\mathbf{E}(\mathbb{F}_q)$ is cyclic itself. We now put

$$\begin{aligned}
\Delta(q) &= \max\{D(\mathbf{E}) \mid \mathbf{E} \in \mathcal{E}(q),\ \mathbf{E}(\mathbb{F}_q) \text{ is cyclic}\} \\
\delta(q) &= \min\{d(\mathbf{E}) \mid \mathbf{E} \in \mathcal{E}(q),\ \mathbf{E}(\mathbb{F}_q) \text{ is cyclic}\}.
\end{aligned}$$

It is shown in [134] that $\Delta(q) \le 7/8$ if q is even, and $\Delta(q) \in \{1/2, 2/3\}$ if q is odd (in fact each case $\Delta(q) = 1/2$ and $\Delta(q) = 2/3$ is fully characterised). Yet another result of [134] asserts that

$$\liminf_{p \to \infty} \delta(p) = 0$$

when $p \to \infty$ runs through the set of primes. Similar quantities are also studied separately for the families of ordinary and supersingular curves.

Vlăduţ [133] has also considered the dual question when the field $\mathbb{F}_q$ is fixed but the curve $\mathbf{E}$ varies over $\mathcal{E}(q)$ (as well as over all ordinary and all supersingular curves from $\mathcal{E}(q)$). In particular, a complete characterisation of prime powers q for which all curves from $\mathcal{E}(q)$ are cyclic is given in Theorem 4.1 of [133]. Such prime powers include, but are not limited to, all *Mersenne primes*. It does not follow immediately from the characterisation given in [133] that there are infinitely many such prime powers, but it looks quite promising that such an unconditional result can be obtained for the prime powers described in the cases (ii) and (iii) of Theorem 4.1 of [133].

Furthermore, Theorem 6.1 of [133] gives an asymptotic formula

$$c(q) = \prod_{\ell \mid q-1} \left(1 - \frac{1}{\ell(\ell^2 - 1)}\right) + O\left(q^{-1/2+o(1)}\right)$$

for the proportion $c(q)$ of the isomorphism classes of curves from $\mathcal{E}(q)$ which contain cyclic curves.

2.2 The Size of the Exponent

For a fixed non-CM elliptic curve $\mathbf{E}$ which is defined over $\mathbb{Q}$, it has been shown by Schoof [121] that

- the inequality

$$\ell_{\mathbf{E}}(p) \gg \frac{p^{1/2} \log p}{\log \log p}$$

holds for all sufficiently large primes p;

- under the ERH, for infinitely many primes,

$$\ell_{\mathbf{E}}(p) \ll p^{7/8} \log p.$$

It is also remarked in [121] that for CM curves the exponent can apparently be small infinitely often. For example, if $\mathbf{E}$ is given by $Y^2 = X^3 - X$, thus $\mathbf{E}$ has complex multiplication over $\mathbb{Z}[i]$, then $\ell_{\mathbf{E}}(p) = k \sim p^{1/2}$ for every prime p of the form $p = k^2 + 1$.

Duke [38] has recently shown, unconditionally for CM elliptic curves and under the ERH for non-CM elliptic curves, that for any function $f(x) \to \infty$ for $x \to \infty$, the lower bound $\ell_{\mathbf{E}}(p) \geq p/f(p)$ holds for almost all primes p. For non-CM elliptic curves, the only unconditional result available is also in [38], and asserts that the weaker inequality $\ell_{\mathbf{E}}(p) \geq p^{3/4}/\log p$ holds for almost all primes p. Moreover, it follows from the proof of that result that in fact for almost all primes p we have $\ell_{\mathbf{E}}(p) \geq p^{3/4}/\log p$ for all elliptic curves $\mathbf{E} \in \mathcal{E}(p)$. In this form, the bound is tight. Indeed, using the *Bombieri-Vinogradov* theorem one can show that for any fixed ε there is a positive proportion of primes p such that $p - 1$ has a prime divisor r with $p^{1/4-\varepsilon} \leq r \leq p^{1/4-\varepsilon/2}$. Thus there is an integer $N \in [p - 2p^{1/2} + 1, p + 2p^{1/2} + 1]$ with $r^2 | N$. Then by [62, 115, 131, 135] there exists an elliptic curve $\mathbf{E} \in \mathcal{E}(p)$ with

$$\mathbf{E}(\mathbb{F}_p) \cong \mathbb{Z}/L\mathbb{Z} \times \mathbb{Z}/M\mathbb{Z},$$

where $M = r$, $L = N/r$ and thus

$$\ell_{\mathbf{E}}(p) = L \ll p^{3/4+\varepsilon},$$

see [45] for more details.

It is useful to note that many of the above results are analogues (albeit somewhat weaker) of the results of [42, 63, 108, 111] on the size of the multiplicative order of a given integer $a > 1$ modulo p, when p varies.

It has also been shown in [121], that, under the ERH, for any curve $\mathbf{E}$ over $\mathbb{Q}$,

$$\liminf_{p\to\infty} \frac{\ell_{\mathbf{E}}(p)}{p^{7/8}\log p} < \infty \tag{2.1}$$

where p runs through all prime numbers. This bound rests on an explicit form of the Chebotarev Density Theorem. Accordingly, unconditional results of [75] lead to an unconditional, albeit much weaker, upper bound on $\ell_{\mathbf{E}}(p)$.

In extension fields of $\mathbb{F}_q$, with $\mathbf{E}$ defined over $\mathbb{F}_q$, stronger lower bounds on $\ell_{\mathbf{E}}(q^n)$ can be obtained. For example, it has recently been shown by Luca and Shparlinski [93] that for any $\varepsilon > 0$, the inequality

$$\ell_{\mathbf{E}}(q^n) \geq q^{n(1-\varepsilon)} \tag{2.2}$$

holds for all but at most $2^{\vartheta\varepsilon^{-6}+O(\varepsilon^{-5})}$ values of n, where $\vartheta = 3^7 2^{-10} = 2.135\ldots$. It also implies that

$$\ell_{\mathbf{E}}(q^n) > q^{n\left(1-\eta(\log n)^{-1/6}\right)}$$

for almost all positive integers n, for any fixed constant

$$\eta > (3^7 2^{-10} \log 2)^{1/6} = 1.067\ldots.$$

The proof of the bound (2.2) is based on recently emerged applications of the celebrated *Subspace Theorem* (see [43, 44] for most recent achievements) as well as on an upper bound due to van der Poorten and Schlickewei [113], on the number of zeros of linear recurrence sequences. It also uses several ideas from [31]. Because the proof of (2.2) is based on the Subspace Theorem the set of exceptional values cannot be effectively determined. Using very different arguments, Luca and Shparlinski [93] have also derived a much weaker but effective bound

$$\ell_{\mathbf{E}}(q^n) \geq q^{n/2+\vartheta(q)n/\log n}$$

which holds for all positive integers n, where $\vartheta(q) > 0$ is an effectively computable constant.

We note that the bound (2.2) means that no result of the same strength as (2.1) is possible for elliptic curves in extension fields.

Accordingly, Luca, McKee and Shparlinski [91] give a more modest bound which asserts that for some absolute constant $\eta > 0$

$$\liminf_{n \to \infty} \frac{\ell_{\mathbf{E}}(q^n)}{q^n \exp\left(-n^{\eta/\log \log n}\right)} < \infty.$$

The proof is based on studying the degree of the field extension of $\mathbb{F}_q$ containing all k-*torsion points*, that is, the points $P \in \mathbf{E}(\overline{\mathbb{F}}_q)$ on $\mathbf{E}$ in the algebraic closure $\overline{\mathbb{F}}_q$ of $\mathbb{F}_q$ with $kP = \mathcal{O}$. It is known that for $\gcd(k, q) = 1$ these points form a group

$$\mathbf{E}[k] \cong \mathbb{Z}/k\mathbb{Z} \times \mathbb{Z}/k\mathbb{Z}$$

of cardinality k^2. Let $d(k)$ denote the degree of the field of definition of $\mathbf{E}[k]$ (that is, the field generated by the coordinates of all the k-torsion points, over $\mathbb{F}_q$). It is shown in [91] that if r is a prime with

$$\gcd\left(r, q\left(a_{\mathbf{E}}(q^n)^2 - 4q^n\right)\right) = 1$$

and such that $a_{\mathbf{E}}(q^n)^2 - 4q^n$ is a quadratic residue modulo r, then $d(r) \mid (r-1)$. Then a modification of a result of [1] is shown which asserts that infinitely many integers n have exponentially many divisors of the form $r - 1$, where r is one of the above primes. For each such n one can easily conclude that $\mathbf{E}(\mathbb{F}_{q^n})$ contains the corresponding torsion subgroups $\mathbf{E}[r]$ which forces $\ell_{\mathbf{E}}(q^n)$ to be sufficiently small compared to q^n.

Most of the proofs are based on the reduction of the original question to a certain number theoretic question about the arithmetic structure of some integers of special form. This reduction is based on one or both of the following two facts:

- By a result of Lenstra [83], for any prime p and any integer N,

$$\#\{\mathbf{E} \in \mathcal{E}(p) \mid N_{\mathbf{E}}(p) = N\} = O(p^{3/2} \log p (\log \log p)^2). \qquad (2.3)$$

- By the Weil pairing, for any prime power e and $\mathbf{E} \in \mathcal{E}(q)$

$$m_{\mathbf{E}}(q) | q - 1, \qquad (2.4)$$

where

$$m_{\mathbf{E}}(q) = N_{\mathbf{E}}(q)/\ell_{\mathbf{E}}(q) \qquad (2.5)$$

Finally, one can also study an apparently easier question about the distribution of $\ell_{\mathbf{E}}(q)$ "on average" over various families of elliptic curves from $\mathcal{E}(q)$, see [125, 133]. For example, it is shown in [125] that "on average" over the curves $\mathbf{E} \in \mathcal{E}(p)$, the value of $m_{\mathbf{E}}(p)$, given by (2.5) is of order at most $\log p (\log \log p)^3$. Indeed, using (2.3) and (2.4), we deduce

$$\sum_{\mathbf{E} \in \mathcal{E}(p)} m_{\mathbf{E}}(p) \leq \sum_{\substack{m | p-1 \\ m \leq p^{1/2}+1}} m \sum_{\substack{|N-p-1| \leq 2p^{1/2} \\ N \equiv 0 \pmod{m^2}}} p^{3/2} \log p (\log \log p)^2$$

$$\leq p^{3/2} \log p (\log \log p)^2 \sum_{\substack{m | p-1 \\ m \leq p^{1/2}+1}} m \left(\frac{4p^{1/2}}{m^2} + 1 \right)$$

$$\ll p^2 \log p (\log \log p)^2 \sum_{m | p-1} \frac{1}{m} \ll p^2 \log p (\log \log p)^3.$$

A related question about the size of the *Tate-Shafarevich group* is considered in [26, 28] which is closely related to the question on the distribution of the largest integer square dividing $a_{\mathbf{E}}(p)^2 - 4p$ where $\mathbf{E}$ is defined over $\mathbb{Q}$ and the prime p varies.

Luca and Shparlinski [94] have studied the dual case where p is fixed and $\mathbf{E}$ runs through the set of all elliptic curves from $\mathcal{E}(p)$ and have shown that for almost all curves the largest square factor of $a_{\mathbf{E}}(p)^2 - 4p$ is logarithmically small.

2.3   Arithmetic Properties of $N_{\mathbf{E}}(q^n)$

It is natural to ask whether $N_{\mathbf{E}}(q)$ can take all values satisfying the inequality (1.2) when $\mathbf{E}$ runs through the set $\mathcal{E}(q)$. This question is fully answered by the classical results of Deuring [36], see also [13, 120, 136]. Namely, for any t with $|t| \leq 2q^{1/2}$, there is a curve $\mathbf{E} \in \mathcal{E}(q)$ with $N_{\mathbf{E}}(q) = q + 1 - t$ if and only if one of the following conditions is satisfied

- $\gcd(t, p) = 1$;

- $q = p^{2m+1}$ is an odd power of p and

  - $t = 0$;
  - $t = \pm(pq)^{1/2}$ for $p = 2$ and $p = 3$;

- $q = p^{2m}$ is an even power of p and

  - $t = 0$, for $p = 2$ and $p \equiv 3 \pmod 4$;
  - $t = \pm q^{1/2}$ for $p = 3$ and $p \equiv 2 \pmod 3$;
  - $t = \pm 2q^{1/2}$.

Lenstra [83] has made it more precise and has shown that for a prime $q = p$ each integer a with $|a| \le 2p^{1/2}$ is taken by $a_{\mathbf{E}}(p)$ for $\mathbf{E} \in \mathcal{E}(p)$ no more than some logarithmic factor of the expected value, namely no more than $O(p^{3/2} \log p (\log \log p)^2)$ times, see (2.3). Moreover in the smaller interval $|a| \le p^{1/2}$, each integer is taken by $a_{\mathbf{E}}(q)$ no less than some logarithmic fraction of the expected value, namely no less than $Cp^{3/2}/\log p$ times, for some absolute constant $C > 0$. Thus, all isogeny classes of elliptic curves contain about the same number of classes of isomorphic curves. These results can also be extended to arbitrary finite fields.

In particular, using the above results, one can immediately find the number of isogeny classes of elliptic curves over $\mathbb{F}_q$. The number $I(q)$ of isomorphism classes of curves from $\mathcal{E}(q)$ is also known:

- if q is odd then

$$I(q) = 2q + 3 + \left(\frac{-1}{q}\right) + 2\left(\frac{-3}{q}\right),$$

  where (a/b) is the Jacobi symbol,

- if q is even then

$$I(q) = 2q + 1,$$

see [120].

Possible group structures which can be realised by $\mathbf{E} \in \mathcal{E}(q)$ are also known, for example, see [62, 115, 131, 135]. Roughly speaking, with only few fully described exceptions, for any integers L and M with $M \mid L$ and such that ML can be realised as a cardinality of an elliptic curve over $\mathbb{F}_q$, there is also $\mathbf{E} \in \mathcal{E}(q)$ for which

$$\mathbf{E}(\mathbb{F}_q) \cong \mathbb{Z}/L\mathbb{Z} \times \mathbb{Z}/M\mathbb{Z}.$$

The dual problem is to study $\pi_{\mathbf{E}}(a, x)$, which is the number of primes for which $a_{\mathbf{E}}(p) = a$ for a given curve $\mathbf{E}$ defined over $\mathbb{Q}$ and a given integer a.

In the case that $\mathbf{E}$ does not have CM or $a \neq 0$, Lang and Trotter [76] conjecture that

$$\pi_{\mathbf{E}}(a, x) \sim c_{\mathbf{E},a} \frac{\sqrt{x}}{\log x}$$

for some constant $c_{\mathbf{E},a}$. In [107], it is shown that

$$\pi_{\mathbf{E}}(a, x) \ll x^{4/5} (\log x)^{-1/5}$$

assuming the ERH for Dedekind zeta functions. For $a = 0$, in [107], one can find a stronger estimate of $O(x^{3/4})$, but this has also been established unconditionally by Elkies [41] using a remark of Murty (see [40]).

Concerning lower bounds, Fouvry and Murty [46] have shown that

$$\pi_{\mathbf{E}}(0, x) \gg \frac{\log \log \log x}{(\log \log \log \log x)^{\delta}}$$

for any $\delta > 0$. Assuming the ERH for classical Dirichlet L-functions, Elkies and Murty (see [40]) have shown that

$$\pi_{\mathbf{E}}(0, x) \gg \log \log x.$$

Fouvry and Murty [46] have also considered $\pi_{\mathbf{E}}(a, x)$ on average over the family of curves $\mathcal{E}(U, V)$ described by all possible Weierstrass equations of the form (1.1) with $|A| \leq U$, $|B| \leq V$. Theorem 6 of [46] gives the expected average order of $\pi_{\mathbf{E}}(a, x)$ whenever $\min\{U, V\} \geq x^{1/2+\varepsilon}$ and $UV \geq x^{3/2+\varepsilon}$ for some fixed $\varepsilon > 0$. This result is based, among other tools, on the *Weil bound* for exponential sums, see Chapter 3 or [86, Chapter 5]. Average values of higher powers of $\pi_{\mathbf{E}}(a, x)$ are estimated by David and Pappalardi [34].

The joint distribution for several curves has been studied by Akbary, David and Juricevic [2], see also [11, 35, 64]. Several more relevant results have recently been obtained by Gekeler [51].

David, Kisilevsky and Pappalardi [33] give a certain characterisation of the pairs $(\mathbf{E}, a)$ of elliptic curves $\mathbf{E}$ over $\mathbf{Q}$ and integers a such that there are no primes p with $a_p(\mathbf{E}) = a$.

It is also relevant to recall that according to the famous *Sato–Tate conjecture*, for every non-CM elliptic curve $\mathbf{E}$ over $\mathbb{Q}$, the ratios $a_{\mathbf{E}}(p)/p^{1/2}$, when p runs through all sufficiently large primes, are distributed in the interval $[-2, 2]$ according to the Sato–Tate measure

$$d\mu_{ST}(x) = \frac{1}{\pi} \sqrt{1 - \frac{x^2}{4}} dx.$$

Studying how often $N_{\mathbf{E}}(q)$ is prime is of great interest, although it appears to be very hard. In both situations when $\mathbf{E}$ is a fixed curve over $\mathbb{Q}$ and p varies and when p is fixed prime and $\mathbf{E}$ runs through $\mathcal{E}(p)$, the question of primality of $N_{\mathbf{E}}(p)$ appears to be out of reach, even if heuristically the situation is well understood (which is quite sufficient for cryptographic applications) thanks to works of Galbraith and McKee [48], Koblitz [69, 72] and Weng [137]. In fact studying primality of $N_{\mathbf{E}}(q^n)/N_{\mathbf{E}}(q)$ when $\mathbf{E}$ is a fixed curve over $\mathbb{F}_q$ and n varies is probably even harder. Clearly, this question is no easier than the question of primality of *Mersenne numbers*.

The only scenario in which rigorous results have been obtained is when both p varies and $\mathbf{E}$ runs through $\mathcal{E}(p)$. In this case Koblitz [70] shows that $N_{\mathbf{E}}(p)$ is prime for the set of pairs $(p, \mathbf{E})$ of the right order of magnitude. The result of [70] follows from the prime number theorem and the aforementioned result of Lenstra [83], which asserts that every integer value, in the interval $[p + 1 - 2p^{1/2}, p + 1 + 2p^{1/2}]$, except maybe for at most two such integers, is taken by $N_{\mathbf{E}}(p)$ about the same number of times when $\mathbf{E}$ runs through $\mathcal{E}(p)$. Moreover, under the ERH there are no exceptions.

We also remark that it follows immediately from the result of Liu and Wu [87] that for a prime power q there is a positive proportion of integers n in the middle part of the Hasse–Weil interval $n \in [q + 1 - q^{1/2}, q + 1 + q^{1/2}]$ with $P(n) \geq n^{0.738}$, where $P(n)$ denotes the greatest prime factor of n. Clearly, each prime number $\ell \geq n^{0.738}$ occurs at most once as $\ell = P(n)$ for $n \in [q + 1 - q^{1/2}, q + 1 + q^{1/2}]$ (provided q is large enough). In fact, the recent result of Harman [58] allows to replace 0.738 with 0.74. Let $K = \lceil q^{1/4}(\log q)^{-1/2} \rceil$. Then removing

$$O\left(\sum_{k=1}^{K} \frac{\log(q^k - 1)}{\log\log(q^k - 1)} \right) = O\left(K^2 \frac{\log q}{\log\log q} \right) = o(q^{1/2})$$

such integers n with

$$P(n) \left| \prod_{k=1}^{K} (q^k - 1) \right.$$

we see that for the other n (which still form a set of positive density) the cardinality of an elliptic curve $\mathbf{E}$ with $N_{\mathbf{E}}(q) = n$ contains a very large prime divisor and also the **MOV** attack [97] (see Section 3.1) has no chance to succeed even if a polynomial time algorithm for the ordinary discrete logarithm problem is discovered.

As a certain "approximation" to the primality question, one may study the values of $\omega(N_{\mathbf{E}}(q))$ and $\Omega(N_{\mathbf{E}}(q))$, where as usual $\omega(m)$ and $\Omega(m)$ are the number of distinct prime factors and the number of prime divisors counted with multiplicity of a positive integer $m > 1$.

Miri and Murty [100] have proved that under the ERH, for any non-CM elliptic curve $\mathbf{E}$ over $\mathbb{Q}$, one has an analogue of the *Turán–Kubilius* inequality:

$$\sum_{p \leq x} \left(\omega \left(N_{\mathbf{E}}(p) \right) - \log \log p \right)^2 = O(\pi(x) \log \log x)$$

where, as usual, $\pi(x)$ is the number of primes $p \leq x$ (see also [27] where the same result is obtained under a weaker hypothesis). For CM curves, a similar result is obtained by Liu [88], see also [89, 90]. In a more general framework, similar questions are considered in [106].

Steuding and Weng [130], improving some previous results of Miri and Murty [100], have shown, under the ERH, that there are at least $C(\mathbf{E})x/(\log x)^2$ primes $p \leq x$ such that $\Omega(N_{\mathbf{E}}(p)) \leq 9$, if $\mathbf{E}$ is a non-CM curve, $\Omega(N_{\mathbf{E}}(p)) \leq 4$, if $\mathbf{E}$ is a CM curve, where the constant $C(\mathbf{E}) > 0$ depends only on $\mathbf{E}$. In the non-CM case, they have also proved that $\omega(N_{\mathbf{E}}(p)) \leq 6$ for the same number of primes $p \leq x$. Note that the above results are actually quoted from the *Erratum* to [130] rather than from the original version claiming slightly stronger estimates. Finally, Cojocaru [27] has obtained an unconditional result, showing that if $\mathbf{E}$ is a CM curve then $\omega(N_{\mathbf{E}}(p)) \leq 5$ for at least $C(\mathbf{E})x/(\log x)^2$ primes $p \leq x$.

A number of other results about the arithmetic structure of $N_{\mathbf{E}}(p)$ (such as divisibility, primality, squarefreeness) and several others associated with $N_{\mathbf{E}}(p)$ quantities are conveniently summarised by Cojocaru [26].

Several more questions on the smoothness of the values of $N_{\mathbf{E}}(p)$, which are also relevant to elliptic curve factorisation [83], have been raised by McKee [96].

3.   Applications to Cryptography

3.1   Menezes-Okamoto-Vanstone Algorithm

The best known general discrete logarithm algorithms in the group $\mathbf{E}(\mathbb{F}_q)$ run in time $Q^{1/2}(\log q)^{1+o(1)}$ where Q is the largest prime divisor of $N_{\mathbf{E}}(q)$ (as for any other abelian group of the same order).

A different algorithm that has been developed for the elliptic curve discrete logarithm is the well-known *Menezes-Okamoto-Vanstone* algorithm, **MOV**, see [97]. This algorithm constructs an embedding of a fixed cyclic subgroup of order L of $\mathbf{E}(\mathbb{F}_p)$ into the multiplicative group $\mathbb{F}_{p^k}^*$ of a suitable extension of $\mathbb{F}_p$, namely such that $L \mid p^k - 1$.

Heuristically, discrete logarithms in $\mathbb{F}_{p^k}^*$ can be found in running time $\mathcal{L}_{p^k}\left(1/3, (64/9)^{1/3}\right)$ by the number field sieve algorithm (see [32, 117, 118]), where, as usual, $\mathcal{L}_m(\alpha, \beta)$ denotes any quantity of the form

$$\mathcal{L}_m(\alpha, \beta) = \exp\left((\beta + o(1))(\log m)^\alpha (\log \log m)^{1-\alpha} \right).$$

In particular, it follows that in order for the running time for **MOV** (combined with the number field sieve) to be subexponential one needs $k \leq (\log p)^2$.

Balasubramanian and Koblitz [4] give an upper bound on the probability that a random pair $(p, \mathbf{E})$ consisting of a prime number p in the interval $[x/2, x]$ and an elliptic curve $\mathbf{E} \in \mathcal{E}(p)$ and having a prime number of points (thus, $N_{\mathbf{E}}(p) = Q$), satisfies the condition that $N_{\mathbf{E}}(p) \mid (p^k - 1)$ for some $k \leq (\log p)^2$. They show that for a sufficiently large x the above probability is $O\left(x^{-1}(\log x)^9 (\log \log x)^2\right)$. This means that for a random elliptic curve with a prime number of points, **MOV** succeeds only with negligible probability.

Luca, Mireles and Shparlinski [92] obtain various extensions of this result. In particular, they do not use the assumption of primality of $N_{\mathbf{E}}(p)$ and thus do not need averaging over primes p in $[x/2, x]$.

Given a curve $\mathbf{E} \in \mathcal{E}(q)$, it would also be interesting to estimate for how many $n \leq x$ **MOV** can be used for the discrete logarithm problem in the groups $\mathbf{E}(\mathbb{F}_{q^n})$. The arguments of [93] can be used to get some results in this direction but they are rather weak.

3.2   Elliptic Curves with Low Embedding Degree

As we have seen in Section 3.1, it is very unlikely for random curve to have a low embedding degree.

On the other hand, since the pioneering works [18, 19, 66, 67, 114, 116, 132] which have introduced several other cryptographic applications of the *Tate or Weil pairing* on elliptic curves (see, for example, [3, 16]), there has been active interest in constructing such rare curves.

In particular, for these applications, the following problem is of primal interest: Find an efficient algorithm to construct an elliptic curve $\mathbf{E} \in \mathcal{E}(q)$, such that $\#\mathbf{E}(\mathbb{F}_q)$ has a sufficiently large prime divisor $\ell \mid N_{\mathbf{E}}(q)$ which also satisfies $\ell \mid q^k - 1$ for a reasonably small value of the positive integer k, see [8–10, 22, 39, 49, 101, 122].

Let

$$\Phi_k(X) = \prod_{\substack{j=0 \\ \gcd(j,k)=1}}^{k} (X - \exp(2\pi \iota j/k))$$

be the kth *cyclotomic polynomial*, where $\iota = \sqrt{-1}$. Typically, the above mentioned constructions work in two steps:

Step 1 Choose a prime ℓ, integers $k \geq 2$ and t, and a prime power q such that

$$|t| \leq 2q^{1/2}, \qquad t \neq 0, 1, 2, \qquad \ell \mid q + 1 - t, \qquad \ell \mid \Phi_k(q). \qquad (3.1)$$

Step 2 Construct an elliptic curve $\mathbf{E} \in \mathcal{E}(q)$ with $N_{\mathbf{E}}(q) = q + 1 - t$ (thus with $a_{\mathbf{E}}(q) = t$).

In the above construction, k should be reasonably small (for example $k = 2, 3, 4, 6$ are typical values), while the ratio $\log \ell / \log q$ should be as large as possible, preferably close to 1.

Unfortunately, there is no efficient algorithm for Step 2, except for the case when $t^2 - 4q$ has a very small square-free part; that is, when

$$t^2 - 4q = -r^2 s \tag{3.2}$$

with some integers r and s, where s is a small square-free positive integer. In this case either s or $4s$ is the fundamental discriminant of the CM field of the corresponding elliptic curve. As we have mentioned in Section 2.2, the result of [94] shows that for every finite field there are very few curves for which the CM discriminant is small. However we now have an additional condition (3.1), which further reduces the number of possible curves. In fact, it has been shown in [95] that there are very few fields for which a suitable elliptic curve may exist.

More precisely, for positive real numbers x, y and z, let us denote by $Q_k(x, y, z)$ the number of prime powers $q \leq x$ for which there exist a prime $\ell \geq y$ and an integer t satisfying (3.1) and (3.2) with a square-free positive integer $s \leq z$. For any fixed integer k, this quantity has been estimated in [95] as

$$Q_k(x, y, z) \leq x^{3/2+o(1)} y^{-1} z, \qquad x \to \infty.$$

In particular, if $z = x^{o(1)}$, which is the only practically interesting case, we conclude that unless $y \leq x^{1/2}$ there are very few finite fields suitable for pairing based cryptography. In other words, unless the request of the primality of $N_{\mathbf{E}}(\mathbb{F}_q)$ is relaxed to the request for $N_{\mathbf{E}}(\mathbb{F}_q)$ to have a large prime divisor (that is, a prime divisor ℓ with $\log \ell / \log q \geq 1/2$), the suitable fields are very rare.

One can also find in [95] a heuristic analysis of one of the first constructions of pairing friendly curves, known as *MNT curves*, see [101], which suggests that this algorithm may produce only finitely many curves over all possible finite fields. On the other hand, it is argued in [95] that the same algorithm may produce infinitely many fields $\mathbb{F}_q$ and curves $\mathbf{E} \in \mathcal{E}(q)$ for which $N_{\mathbf{E}}(q)$ has a prime divisor $\ell | N_{\mathbf{E}}(q)$ of size $\ell = q^{1+o(1)}$.

3.3   Distribution of Points and Pseudorandom Number Generators

Given a curve $\mathbf{E} \in \mathcal{E}(p)$ it is natural to ask how (visually) the set of points of $\mathbf{E}(\mathbb{F}_p)$ looks like, that is, if one plots all finite points $P \in \mathbf{E}(\mathbb{F}_p)$ inside of the square $[0, p-1] \times [0, p-1]$, how does this picture look like?

We start with an observation that the traditional picture, which occurs in almost all books on elliptic curves, illustrating an elliptic curve as an oval and a line has nothing to do with how elliptic curves over finite fields look like. In fact, the bound of Bombieri [17] immediately implies that the points $P \in \mathbf{E}(\mathbb{F}_p)$ are uniformly distributed in $[0, p-1] \times [0, p-1]$, thus the picture is a rather unexciting uniformly grey square.

Kohel and Shparlinski [74] have proved that this picture does not change much even if one considers only points from an arbitrary sufficiently large subgroup **H** of $\mathbf{E}(\mathbb{F}_p)$, namely of any subgroup of size $\#\mathbf{H} \geq p^{1/2+\varepsilon}$, where p is a sufficiently large prime. This result is based on the following estimate of exponential sums

$$\sum_{P \in \mathbf{H}} \exp\left(2\pi i f(P)/p\right) = O(p^{1/2}), \tag{3.3}$$

which holds for any subgroup $\mathbf{H} \in \mathbf{E}(\mathbb{F}_p)$ and any function f which is not constant on **E** (and a similar bound for additive character sums in extension fields). Beelen and Doumen [12] give some analogues of the results of [74], including bounds of sums of multiplicative characters over points of elliptic curves.

Hess and Shparlinski [60] have used (3.3) to study the uniformity of distribution properties of the elliptic curve analogue of the *linear congruential generator* which is a sequence of points $Q_n \in \mathbf{E}(\mathbb{F}_q)$ satisfying the relation

$$Q_n = Q_{n-1} + G = nG + Q_0, \qquad n = 1, 2, \dots.$$

where $Q_0 \in \mathbf{E}(\mathbb{F}_q)$ is initial value and $G \in \mathbf{E}(\mathbb{F}_q)$ is a fixed point of sufficiently large order.

Lange and Shparlinski [80] have estimated exponential sums along the sequence of points $P_n \in \mathbf{E}(\mathbb{F}_q)$ satisfying the relation

$$P_n = eP_{n-1} = e^n G, \qquad n = 1, 2, \dots.$$

for some fixed integer e and the initial value $P_0 = G$ (see also Chapter 4, Section 2.1). In fact this sequence gives an elliptic curve analogue of the so-called *power generator* of pseudorandom numbers.

The bound (3.3) has also been applied in [124] to study the distribution of the *Naor-Reingold generator* on elliptic curves, which for a given point $G \in \mathbf{E}(\mathbb{F}_p)$ and a k-dimensional integer vector $\mathbf{a} = (a_1, \dots a_k)$, is defined as the sequence:

$$F_{\mathbf{a}}(n) = a_1^{\nu_1} \dots a_k^{\nu_k} G, \qquad n = 1, 2, \dots,$$

where $n = \nu_1 \dots \nu_k$ is the bit representation of n, $0 \leq n \leq 2^k - 1$.

Hess and Shparlinski [60], Lange and Shparlinski [80], Shparlinski and Silverman [126] study some structural properties of various sequences of points on elliptic curves which are relevant to applications to pseudorandom numbers.

In particular, establishing these properties is of independent interest and related to the notion of *linear complexity* (see Chapter 4). However, they are also crucial in order to be able to apply the bound of [74] and prove corresponding uniformity of distribution results in [60, 80, 124].

Recently, Banks, Friedlander, Garaev and Shparlinski [6], have shown that for any prime p and $\mathbf{E} \in \mathcal{E}(p)$

$$\max_{a \in \mathbb{F}_p^*} \left| \sum_{u \in \mathcal{U}} \sum_{v \in \mathcal{V}} \alpha_u \beta_v \exp(2\pi i a \mathbf{x}(uvG)/p) \right| \ll ABt^{5/6}(\#\mathcal{U}\#\mathcal{V})^{1/2}p^{1/12+\varepsilon},$$

where $G \in \mathbf{E}(\mathbb{F}_p)$ is of order t, $\mathbf{x}(Q)$ denotes the x-coordinate of $Q = (\mathbf{x}(Q), \mathbf{y}(Q)) \in \mathbf{E}(\mathbb{F}_p)$, $\mathcal{U}$ and $\mathcal{V}$ are arbitrary subsets of $\mathbb{Z}/t\mathbb{Z}$, α_u and β_v are arbitrary real numbers with

$$A = \max_{u \in \mathcal{U}} |\alpha_u|, \qquad B = \max_{v \in \mathcal{V}} |\beta_v|,$$

and the implied constant depends only on arbitrary $\varepsilon > 0$. Such bounds can be used to study the distribution of the *Diffie-Hellman triples* (uG, vG, uvG) over elliptic curves and also have many other applications. For example, they can be applied to estimate the distribution of points rG, where r runs through a sequence of prime numbers or a sequence of smooth numbers, see [6] for more details.

3.4 Fast Generation of Points

Given a curve $\mathbf{E} \in \mathcal{E}(q)$ and a point $P = (x, y) \in \mathbf{E}(\mathbb{F}_{q^n})$ one can immediately obtain n more (usually distinct) points by applying the *Frobenius endomorphism* φ, which operates on the points by raising each coordinate to the power of q:

$$\varphi^i(P) = (x^{q^i}, y^{q^i}), \qquad i = 0, \ldots, n-1.$$

One can now consider linear combinations

$$\sum_{i=0}^{n-1} m_i \varphi^i(P), \qquad (m_0, \ldots, m_{n-1}) \in \mathcal{M}, \qquad (3.4)$$

where the set $\mathcal{M}$ is such that the above linear combinations are easy to compute. A natural example would be a set of binary vectors. However, a cryptographically more interesting example is given by the set $\mathcal{M}$ of *non-adjacent* $0, \pm 1$ vectors. That is, the set of vectors having only $0, \pm 1$ as components and not having two consecutive nonzero components. It also follows from Proposition 4 of [21] that for $n \geq 2$ we have $\#\mathcal{M} = 2^{n+2}/3 + O(1)$.

Solinas [129] gives a detailed study of the above set of points in the case of the *Koblitz curves* [71]

$$Y^2 + XY = X^3 + aX^2 + 1, \qquad a = 0, 1,$$

defined over $\mathbb{F}_2$. Various generalisations, including generalisations to hyperelliptic curves have been proposed by Günther, Lange and Stein [54] and Lange [78, 79].

It is certainly not guaranteed that points (3.4) are pairwise distinct (and in fact this is generally not true). However, Lange and Shparlinski [81] have obtained a tight upper bound on the number of repetitions in some interesting and practically important cases, such as the Koblitz curves with the above set of non-adjacent vectors. Several results about the uniformity of distribution of such and similar points have recently been obtained in [82].

3.5 Compact Weierstrass Equations

Assume that $\mathbf{E}$ is given by a Weierstrass equation (1.1) over $\mathbb{F}_p$ (with $p \geq 5$). It is certainly natural to ask what is the smallest value $H_{\mathbf{E}}(p)$ for which one can find a possibly different pair of coefficients A and B with $0 \leq A, B \leq H_{\mathbf{E}}(p)$ for which the corresponding equation defines an elliptic curve isomorphic to $\mathbf{E}$.

The question has been introduced by Ciet, Quisquater and Sica in [23], where the bound $H_{\mathbf{E}}(p) = O(p^{3/4})$ is given. It has also been shown by Banks and Shparlinski [7] that for almost all curves from $\mathcal{E}(p)$, one has $H_{\mathbf{E}}(p) = O(p^{2/3})$. These results are based on using exponential sums to show that the residues of At^4 and Bt^6 modulo p can be made simultaneously small for some integer $t \in \mathbb{F}_p^*$. This is required for all pairs $(A, B) \in \mathbb{F}_p^2$ for the purposes of [23], and for almost all pairs $(A, B) \in \mathbb{F}_p^2$ for the purposes of [7].

In fact the method of proof used in [23] is almost identical to that used in [46]. So it is natural to try to combine the arguments of [46] with those of [7] and improve the result of [46]. However this, if possible at all, seems to require bringing in some new ideas.

3.6 Computation of the Group Structure

The algorithm of R. Schoof [119] computes $N_{\mathbf{E}}(q)$ in deterministic polynomial time, see also [3, 15, 16, 50] for more recent improvements (both theoretic and practical). However, computing the group structure (1.3) seems to be much more complicated.

The deterministic algorithm of [74] computes the group structure of any elliptic curve $\mathbf{E} \in \mathcal{E}(q)$ in exponential time $O(q^{1/2+o(1)})$ which is too slow for practical applications.

A more efficient but probabilistic algorithm of Miller [99] runs in expected polynomial time plus the time needed to factor $\gcd(N_{\mathbf{E}}(q), q-1)$, see also [14]. Friedlander, Pomerance and Shparlinski [47] have shown that for a sufficiently large p and for almost all elliptic curves $\mathbf{E} \in \mathcal{E}(p)$, the factorisation part of the algorithm is in fact less time consuming than the rest of the computation (since $\gcd(N_{\mathbf{E}}(p), p - 1)$ tends to be rather small).

3.7   Primality Testing

It is well known that elliptic curves play an important role in modern factorisation and primality proving algorithms, see [32]. It is probably less known that they can be used to design an analogue of the *Fermat primality test*. Namely given a point P on a CM curve $\mathbf{E}$ over $\mathbb{Q}$, one can verify that for any sufficiently large prime $p \equiv 3 \pmod 4$ the point $(p + 1)P$ is the point at infinity on the reduction of $\mathbf{E}$ modulo p. Accordingly, a composite $n \equiv 3 \pmod 4$ is called an *elliptic pseudoprime* (with respect to the curve $\mathbf{E}$ and the point P) if a similar property holds modulo n. It has been shown in [5, 52, 53, 102] that elliptic pseudoprimes form a sparse set, so this test may supplement other primality tests.

References

[1] L. M. Adleman, C. Pomerance and R. S. Rumely, "On distinguishing prime numbers from composite numbers", Annals Math, Vol. 117, 173–206 (1983).

[2] A. Akbary, C. David and R. Juricevic, "Average distributions and product of L-series", Acta Arith., Vol. 111, 239–268 (2004).

[3] R. Avanzi, H. Cohen, C. Doche, G. Frey, T. Lange and K. Nguyen, *Elliptic and hyperelliptic curve crytography: Theory and practice*, CRC Press, 2005.

[4] R. Balasubramanian and N. Koblitz, "The improbability that an elliptic curve has subexponential discrete log problem under the Menezes-Okamoto-Vanstone algorithm", J. Cryptology, Vol. 11, 141–145 (1998).

[5] R. Balasubramanian and M. R. Murty, "Elliptic pseudoprimes", *Sémin. Théor. Nombres, Paris 1988-89*, Birkhäuser, Boston, MA, Prog. Math., Vol. 91, 13–25 (1990).

[6] W. D. Banks, J. B. Friedlander, M. Garaev and I. E. Shparlinski, "Double character sums over elliptic curves and finite fields", Pure and Appl. Math. Quart., Vol. 2., 179–197 (2006).

[7] W. Banks and I. E. Shparlinski, "Average normalizations of elliptic curves", Bull. Austral. Math. Soc., Vol. 66, 353–358 (2002).

[8] P. S. L. M. Barreto, B. Lynn and M. Scott, "Elliptic curves with prescribed embedding degrees", Lect. Notes in Comp. Sci., Vol. 3006, 17–25 (2003).

[9] P. S. L. M. Barreto, B. Lynn and M. Scott, ' 'Efficient implementation of pairing-based cryptosystems", J. Cryptology, Vol. 17, 297–319 (2004).

[10] P. S. L. M. Barreto and M. Naehrig, "Pairing-friendly elliptic curves of prime order", Lect. Notes in Comp. Sci., Vol. 3897, 319–331 (2006).

[11] J. Battista, J. Bayless, D. Ivanov and K. James, "Average Frobenius distributions for elliptic curves with nontrivial rational torsion", Acta Arith., Vol. 119, 81–91 (2005).

[12] P. Beelen and J. Doumen, "Pseudorandom sequences from elliptic curves", *Finite Fields with Applications to Coding Theory, Cryptography and Related Areas*, Springer-Verlag, Berlin, 37–52 (2002).

[13] B. J. Birch, "How the number of points of an elliptic curve over a fixed prime field varies", J. Lond. Math. Soc., Vol. 43, 57–60 (1968).

[14] I. F. Blake, V. K. Murty and G. Xu, "Refinements of Miller's algorithm for computing the Weil/Tate pairing", J. Algorithms, Vol. 58, 134–149 (2006).

[15] I. F. Blake, G. Seroussi and N. Smart, *Elliptic curves in cryptography*, London Math. Soc., Lecture Note Series, Vol. 265, Cambridge Univ. Press, 1999.

[16] I. F. Blake, G. Seroussi and N. Smart, *Advances in elliptic curves in cryptography*, London Math. Soc., Lecture Note Series, Vol. 317, Cambridge Univ. Press, 2005.

[17] E. Bombieri, "On exponential sums in finite fields", Amer. J. Math., Vol. 88, 71–105 (1966).

[18] D. Boneh and M. Franklin, "Identity-based encryption from the Weil pairing", SIAM J. Comp., Vol. 32, 586–615 (2003).

[19] D. Boneh, B. Lynn and H. Shacham, "Short signatures from the Weil pairing", J. Cryptology, Vol. 17, 297–319 (2004).

[20] I. Borosh, C.J. Moreno and H. Porta, "Elliptic curves over finite fields, II", Math. Comp., Vol. 29, 951-964 (1975).

[21] W. Bosma, "Signed bits and fast exponentiation", J. Théorie des Nombres Bordeaux, Vol. 13, 27–41 (2001).

[22] F. Brezing and A. Weng, "Elliptic curves suitable for pairing based cryptography", Designs, Codes and Cryptography, Vol. 37, 133–141 (2005).

[23] M. Ciet, J.-J. Quisquater and F. Sica, "Elliptic curve normalization", *Crypto Group Technical Report Series CG-2001/2*, Univ. Catholique de Louvain, Belgium, 1–13 (2001).

[24] A. Cojocaru, "On the cyclicity of the group of $\mathbb{F}_p$-rational points of non-CM elliptic curves", J. Number Theory, Vol. 96, 335–350 (2002).

[25] A. Cojocaru, "Cyclicity of CM elliptic curves modulo p", Trans. Amer. Math. Soc., Vol. 355, 2651–2662 (2003).

[26] A. Cojocaru, "Questions about the reductions modulo primes of an elliptic curve", *Proc. 7th Meeting of the Canadian Number Theory Association (Montreal, 2002)*, CRM Proceedings and Lecture Notes, Vol. 36, Amer. Math. Soc., 61–79 (2004).

[27] A. Cojocaru, "Reductions of an elliptic curve with almost prime orders", Acta Arith., Vol. 119, 265–289 (2005).

[28] A. Cojocaru and W. Duke, "Reductions of an elliptic curve and their Tate-Shafarevich groups", Math. Annalen, Vol. 329, 513–534 (2004).

[29] A. Cojocaru, E. Fouvry and M. R. Murty, "The square sieve and the Lang–Trotter conjecture", Canadian J. Math., Vol 57, 1155-1177, (2005).

[30] A. Cojocaru and M. R. Murty, "Cyclicity of elliptic curves modulo p and elliptic curve analogues of Linnik's problem", Math. Annalen, Vol. 330, 601–625 (2004).

[31] P. Corvaja and U. Zannier, "A lower bound for the height of a rational function at S-unit points", Monatsh. Math., Vol. 144, 203–224 (2005).

[32] R. Crandall and C. Pomerance, *Prime numbers: A computational perspective*, Springer-Verlag, New York, 2005.

[33] C. David, H. Kisilevsky and F. Pappalardi, "Galois representations with non-surjective traces", Canad. J. Math., Vol. 51, 936–951 (1999).

[34] C. David and F. Pappalardi, "Average Frobenius distribution of elliptic curves", Internat. Math. Res. Notices, Vol. 4, 165–183 (1999).

[35] C. David and F. Pappalardi, "Average Frobenius Distribution for inerts in $\mathbb{Q}(i)$", J. Ramanujan Math. Soc., Vol. 19, 1–21 (2004).

[36] M. Deuring, "Die Typen der Multiplikatorenringe elliptischer Funktionenkörper", Abh. Math. Sem. Hansischen Univ., Vol. 14, 197–272 (1941).

[37] C. Doche, K. Ford and I. E. Shparlinski, "On finite fields with Jacobians of small exponent", preprint, 2005.

[38] W. Duke, "Almost all reductions modulo p of an elliptic curve have a large exponent", Comptes Rendus Mathematique, Vol. 337, 689–692 (2003).

[39] R. Dupont, A. Enge and A. Morain, "Building curves with arbitrary small MOV degree over finite prime fields", J. Cryptology, Vol. 18, 79–89 (2005).

[40] N. Elkies, *Supersingular primes of a given elliptic curve over a number field*, PhD thesis, Harvard University, 1987.

[41] N. Elkies, "Distribution of supersingular primes", Astérisque, No. 198-200, 127–132 (1991).

[42] P. Erdös and R. Murty, "On the order of a (mod p)", *Proc. 5th Canadian Number Theory Association Conf.*, Amer. Math. Soc., Providence, RI, 87–97 (1999).

[43] J.-H. Evertse, "An improvement of the quantitative subspace theorem", Compos. Math., Vol. 101, 225–311 (1996).

[44] J.-H. Evertse and H. P. Schlickewei, "A quantitative version of the absolute subspace theorem", J. Reine Angew. Math., Vol. 548, 21–127 (2002).

[45] K. Ford and I. E. Shparlinski, "On finite fields with Jacobians of small exponent", preprint, 2005.

[46] E. Fouvry and M. R. Murty, "On the distribution of supersingular primes", Canad. J. Math., Vol. 48, 81–104 (1996).

[47] J. B. Friedlander, C. Pomerance and I. E. Shparlinski, "Finding the group structure of elliptic curves over finite fields", Bull. Aust. Math. Soc., Vol. 72, 251–263 (2005).

[48] S. D. Galbraith and J. McKee, "The probability that the number of points on an elliptic curve over a finite field is prime", J. London Math. Soc., Vol. 62, 671–684 (2000).

[49] S. D. Galbraith, J. McKee and P. Valenca, "Ordinary abelian varieties having small embedding degree", *Proc. Workshop on Math. Problems and Techniques in Cryptology*, CRM, Barcelona, 29–45 (2005).

[50] S. D. Galbraith and A. Menezes, "Algebraic curves and cryptography", Finite Fields and Their Appl., Vol. 11, 544–577 (2005).

[51] E.-U. Gekeler, "Frobenius distributions of elliptic curves over finite prime fields", Int. Math. Res. Notes, Vol. 2003, 1999–2018 (2003).

[52] D. M. Gordon, "On the number of elliptic pseudoprimes", Math. Comp., Vol. 52, 231–245 (1989).

[53] D. Gordon and C. Pomerance, "The distribution of Lucas and elliptic pseudoprimes", Math. Comp., Vol. 57, 825–838 (1991).

[54] C. Günther, T. Lange and A. Stein, "Speeding up the arithmetic on Koblitz curves of genus two", Lect. Notes in Comp. Sci., Vol. 2012, 106–117 (2001).

[55] R. Gupta and M. R. Murty, "A remark on Artin's conjecture", Invent. Math., Vol. 78, 127–130 (1984).

[56] R. Gupta and M. R. Murty, "Primitive points on elliptic curves", Compos. Math. Vol. 58, 13–44 (1986).

[57] R. Gupta and M. R. Murty, "Cyclicity and generation of points   mod p on elliptic curves", Invent. Math., Vol. 101, 225–235 (1990).

[58] G. Harman, *Prime-detecting sieves*, Princeton Univ. Press, Princeton, NY, to appear.

[59] D. R. Heath-Brown, "Artin's conjecture for primitive roots", Quart. J. Math. Vol. 37, 27–38 (1986).

[60] F. Hess and I. E. Shparlinski, "On the linear complexity and multidimensional distribution of congruential generators over elliptic curves", Designs, Codes and Cryptography, Vol. 35, 111–117 (2005).

[61] C. Hooley, "On Artin's conjecture", J. Reine Angew. Math., Vol. 225, 209–220 (1967).

[62] E. W. Howe, "On the group orders of elliptic curves over finite fields", Compositio Math., Vol. 85, 229–247 (1993).

[63] H.-K. Indlekofer and N. M. Timofeev, "Divisors of shifted primes", Publ. Math. Debrecen, Vol. 60, 307–345 (2002).

[64] K. James, "Average Frobenius distributions for elliptic curves with 3-torsion", J. Number Theory, Vol. 109, 278–298 (2004).

[65] E. Jensen and M. R. Murty, "Artin's conjecture for polynomials over finite fields", *Number Theory*, Birkhäuser, Basel, 167–181 (2000).

[66] A. Joux, "A one round protocol for tripartite Diffie–Hellman", Lect. Notes in Comp. Sci., Vol. 1838, 385–393 (2000).

[67] A. Joux, "The Weil and Tate pairings as building blocks for public key cryptosystems", Lect. Notes in Comp. Sci., Vol. 2369, 20–32 (2002).

[68] N. Koblitz, "Elliptic curve cryptosystems", Math. Comp., Vol. 48, 203–209 (1987).

[69] N. Koblitz, "Primality of the number of points on an elliptic curve over a finite field", Pacific J. Math., Vol. 131, 157–166 (1988).

[70] N. Koblitz, "Elliptic curve implementation of zero-knowledge blobs", J. Cryptology, Vol. 4, 207–213 (1991).

[71] N. Koblitz, "CM curves with good cryptographic properties", Lect. Notes in Comp. Sci., Vol. 576, 279–287 (1992).

[72] N. Koblitz, "Almost primality of group orders of elliptic curves defined over small finite fields", Experiment. Math., Vol. 10, 553–558 (2001).

[73] N. Koblitz, "Good and bad uses of elliptic curves in cryptography", Moscow Math. J., Vol. 2, 693–715 (2002).

[74] D. R. Kohel and I. E. Shparlinski, "Exponential sums and group generators for elliptic curves over finite fields", Lect. Notes in Comp. Sci., Vol. 1838, 395–404 (2000).

[75] J. C. Lagarias, H. L. Montgomery and A. M. Odlyzko, "A bound for the least prime ideal in the Chebotarev density theorem", Invent. Math., Vol. 54, 271–296 (1979).

[76] S. Lang and H. Trotter, *Frobenius distributions in GL_2 extensions*, Lecture Notes in Mathematics, Vol. 504, 1976.

[77] S. Lang and H. Trotter, "Primitive points on elliptic curves", Bull. Amer. Math. Soc., Vol. 83, 289–292 (1977).

[78] T. Lange, *Efficient arithmetic on hyperelliptic curves*, PhD thesis, Universität Gesamthochschule Essen, 2001.

[79] T. Lange, "Koblitz curve cryptosystems", Finite Fields and Their Appl., Vol. 11, 200–229 (2005).

[80] T. Lange and I. E. Shparlinski, "Certain exponential sums and random walks on elliptic curves", Canad. J. Math., Vol. 57, 338–350 (2005).

[81] T. Lange and I. E. Shparlinski, "Collisions in fast generation of ideal classes and points on hyperelliptic and elliptic curves", Appl. Algebra in Engin., Commun. and Computing, Vol. 15, 329–337 (2005).

[82] T. Lange and I. E. Shparlinski, "Distribution of some sequences of points on elliptic curves", preprint, 2006.

[83] H. W. Lenstra, "Factoring integers with elliptic curves", Ann. Math., Vol. 126, 649–673 (1987).

[84] S. Li and C. Pomerance, "Primitive roots: A survey", *Number Theoretic Methods (Iizuka, 2001)*, Kluwer Acad. Publ., Dordrecht, 219–231 (2002).

[85] S. Li and C. Pomerance, "On generalizing Artin's conjecture on primitive roots to composite moduli", J. Reine Angew. Math., Vol. 556, 205–224 (2003).

[86] R. Lidl and H. Niederreiter, *Finite fields*, Cambridge University Press, Cambridge, 1997.

[87] H.-Q. Liu and J. Wu, "Numbers with a large prime factor", Acta Arith., Vol. 89, 163–187 (1999).

[88] Y.-R. Liu, "Prime divisors of the number of rational points on elliptic curves with complex multiplication", Bull. London Math. Soc., Vol. 37, 658–664 (2005).

[89] Y.-R. Liu, "A prime analogue to Erdős-Pomerance's conjecture for elliptic curves", Comment. Math. Helv., Vol. 80, 755–769 (2005).

[90] Y.-R. Liu, "Prime analogues of the Erdős-Kac theorem for elliptic curves", J. Number Theory, to appear.

[91] F. Luca, J. McKee and I. E. Shparlinski, "Small exponent point groups on elliptic curves", J. Théorie des Nombres Bordeaux, to appear.

[92] F. Luca, D. J. Mireles and I. E. Shparlinski, "MOV attack in various subgroups on elliptic curves", Illinois J. Math., Vol. 48, 1041–1052 (2004).

[93] F. Luca and I. E. Shparlinski, "On the exponent of the group of points on elliptic curves in extension fields", Intern. Math. Research Notices, Vol. 2005, 1391–1409 (2005).

[94] F. Luca and I. E. Shparlinski, "Discriminants of complex multiplication fields of elliptic curves over finite fields", preprint, 2005.

[95] F. Luca and I. E. Shparlinski, "Elliptic curves with low embedding degree", preprint, 2005.

[96] J. McKee, "Subtleties in the distribution of the numbers of points on elliptic curves over a finite prime field", J. London Math. Soc., Vol. 59, 448–460 (1999).

[97] A. Menezes, T. Okamoto and S. A. Vanstone, "Reducing elliptic curve logarithms to logarithms in a finite field", IEEE Transactions on Information Theory, Vol. 39, 1639–1646 (1993).

[98] V. S. Miller, "Uses of elliptic curves in cryptography", Lect. Notes in Comp. Sci., Vol. 218, 417–426 (1986).

[99] V. S. Miller, "The Weil pairing and its efficient calculation", J. Cryptology, Vol. 17, 235–261 (2004).

[100] S. A. Miri and V. K. Murty, "An application of sieve methods to elliptic curves", Lect. Notes in Comp. Sci., Vol. 2247, 91–98 (2001).

[101] A. Miyaji, M. Nakabayashi and S. Takano, "New explicit conditions of elliptic curve traces for FR-reduction", IEICE Trans. Fundamentals, Vol. E84-A, 1234–1243 (2001).

[102] I. Miyamoto and M. R. Murty, "Elliptic pseudoprimes", Math. Comp., Vol. 53, 415–430 (1989).

[103] M. R. Murty, "On Artin's conjecture", J. Number Theory, Vol. 16, 147–168 (1983).

[104] M. R. Murty, "An analogue of Artin's conjecture for abelian extensions", J. Number Theory, Vol. 18, 241–248 (1984).

[105] M. R. Murty, "Artin's conjecture and elliptic analogues", *Sieve Methods, Exponential Sums, and their Applications in Number Theory*, Cambridge Univ. Press., 325–344 (1996).

[106] M. R. Murty and V. K. Murty, "Prime divisors of Fourier coefficients of modular forms", Duke Math. J., Vol. 51, 57–76 (1984).

[107] M. R. Murty, V. K. Murty and N. Saradha, "Modular forms and the Chebotarev density theorem", American J. Math., Vol. 110, 253-281 (1988).

[108] M. R. Murty, M. Rosen and J. H. Silverman, "Variations on a theme of Romanoff", Intern. J. Math. Soc., Vol. 7, 373–391 (1996).

[109] M. R. Murty and S. Srinivasan, "Some remarks on Artin's conjecture", Canad. Math. Bull., Vol. 30, 80–85 (1987).

[110] W. Narkiewicz, *Classical problems in number theory*, Polish Sci. Publ., Warszawa, 1986.

[111] F. Pappalardi, "On the order of finitely generated subgroups of $\mathbb{Q}^*$ (mod p) and divisors of $p - 1$", J. Number Theory, Vol. 57, 207–222 (1996).

[112] F. Pappalardi and I. E. Shparlinski, "On Artin's conjecture over function fields", Finite Fields and Their Appl., Vol. 1, 399–404 (1995).

[113] A. J. van der Poorten and H. P. Schlickewei, "Zeros of recurrence sequences", Bull. Austral Math. Soc., Vol. 44, 215–223 (1991).

[114] K. Rubin and A. Silverberg, "Supersingular abelian varieties in cryptology", Lect. Notes in Comp. Sci., Vol. 2442, 336–353 (2002).

[115] H.-G. Rück, "A note on elliptic curves over finite fields", Math. Comp., Vol. 49, 301–304 (1987).

[116] R. Sakai, K. Ohgishi and M. Kasahara, "Cryptosystems based on pairing", *Proc. of SCIS'2000*, Okinawa, Japan, 2000.

[117] O. Schirokauer, "Discrete logarithms and local units", Philos. Trans. Roy. Soc. London, Ser. A, Vol. 345, 409–423 (1993).

[118] O. Schirokauer, D. Weber and T. Denny, "Discrete logarithms: The effectiveness of the index calculus method", Lect. Notes in Comp. Sci., Vol. 1122, 337–362 (1996).

[119] R. Schoof, "Elliptic curves over finite fields and the computation of square roots mod p", Math. of Comp., Vol. 44, 483–494 (1985).

[120] R. Schoof, "Nonsingular plane cubic curves over finite fields", J. Combin. Theory, Ser. A, Vol. 47, 183–211 (1987).

[121] R. Schoof, "The exponents of the group of points on the reduction of an elliptic curve", *Arithmetic Algebraic Geometry*, Progr. Math., Vol. 89, Birkhäuser, Boston, MA, 325–335 (1991).

[122] M. Scott and P. S. L. M. Barreto, "Generating more MNT elliptic curves", Designs, Codes and Cryptography, to appear.

[123] J.-P. Serre, "Résumé des cours de 1977-1978", Collected Papers, Vol. III, Springer Verlag, Berlin, 465–468 (1986).

[124] I. E. Shparlinski, "On the Naor–Reingold pseudo-random function from elliptic curves", Appl. Algebra in Engin., Commun. and Computing, Vol. 11, 27–34 (2000).

[125] I. E. Shparlinski, "Orders of points on elliptic curves", *Affine Algebraic Geometry*, Amer. Math. Soc., 245–252 (2005).

[126] I. E. Shparlinski and J. H. Silverman, "On the linear complexity of the Naor–Reingold pseudo-random function from elliptic curves", Designs, Codes and Cryptography, Vol. 24, 279–289 (2001).

[127] J. H. Silverman, *The arithmetic of elliptic curves*, Springer-Verlag, Berlin, 1995.

[128] J. H. Silverman and J. Tate, *Rational points on elliptic curves*, Springer-Verlag, Berlin, 1992.

[129] J. Solinas, "Efficient arithmetic on Koblitz curves", Designs, Codes and Cryptography, Vol. 19, 195–249 (2000).

[130] J. Steuding and A. Weng, "On the number of prime divisors of the order of elliptic curves modulo p", Acta Arith., Vol. 117, 341–352 (2005).

[131] M. Tsfasman and S. Vlăduţ, *Algebraic-Geometric Codes*, Kluwer Acad. Pres, Dordrecht, 1991.

[132] E. R. Verheul, "Evidence that XTR is more secure than supersingular elliptic curve cryptosystems", Lect. Notes in Comp. Sci., Vol. 2045, 195–210 (2001).

[133] S. G. Vlăduţ, "Cyclicity statistics for elliptic curves over finite fields", Finite Fields and Their Appl., Vol. 5, 13–25 (1999).

[134] S. G. Vlăduţ, "On the cyclicity of elliptic curves over finite field extensions", Finite Fields and Their Appl., Vol. 5, 354–363 (1999).

[135] J.F. Voloch, "A note on elliptic curves over finite fields", Bull. Soc. Math. Franc., Vol. 116, 455–458 (1988).

[136] W. C. Waterhouse, "Abelian varieties over finite fields", Ann. Sci. Ecole Norm. Sup., Vol. 2, 521–560 (1969).

[137] A. Weng, "On group orders of rational points of elliptic curves", Quaest. Math., Vol. 25, 513–525 (2002).

Appendix: Algebraic Function Fields

In large parts of this book, the basic theory of algebraic function fields is assumed. In this appendix we collect the main definitions, notations and results of this theory. For a detailed exposition the reader is referred to the books "Algebraic Function Fields and Codes" by H. Stichtenoth (Springer Universitext, 1993) and "Rational Points on Curves over Finite Fields" by H. Niederreiter and C. P. Xing (London Math. Soc. Lecture Notes Ser. **285**, 2001).

(1) An *algebraic function field* F/K is a finite field extension of the rational function field $K(x)$ where K is a perfect field. We always assume implicitely that the field K is algebraically closed in F (i.e.; every element $z \in F$ which is algebraic over K is already in K). The field K is called the *constant field* of F. We consider in this book mostly function fields $F/\mathbb{F}_q$ where $\mathbb{F}_q$ is the finite field with q elements. Such function fields are also called *global function fields*.

(2) A *place* of F is, by definition, the maximal ideal of some valuation ring $\mathcal{O}$ of F/K. To every place P there corresponds a unique normalized *discrete valuation*, denoted v_P or ν_P, which is a surjective map from F to $\mathbb{Z} \cup \{\infty\}$ satisfying the following properties:
   (i) $v_P(x) = \infty$ if and only if $x = 0$.
   (ii) $v_P(xy) = v_P(x) + v_P(y)$ for all $x, y \in F$.
   (iii) $v_P(x + y) \geq \min(v_P(x), v_P(y))$ for all $x, y \in F$.
   (iv) $v_P(a) = 0$ for all $a \in K^{\times}$.
In terms of the valuation v_P, the corresponding valuation ring $\mathcal{O} = \mathcal{O}_P$ of the place P is then given as $\mathcal{O}_P = \{x \in F \mid v_P(x) \geq 0\}$, and the place P is given as $P = \{x \in F \mid v_P(x) > 0\}$. The residue class field $\mathcal{O}_P/P$ is a finite extension of the constant field K, and the *degree of the place* P is defined as

$$\deg P = [\mathcal{O}_P/P : K].$$

195

The place P is said to be a *rational place* if $\deg P = 1$. In this case we have the residue class map at P as follows:

$$\mathcal{O}_P \to K, \quad f \mapsto f(P),$$

where $f(P) \in K$ is the residue class of f in $K = \mathcal{O}_P/P$.

(3) A *divisor* D of F/K is a formal sum $D = \sum_P a_P P$ of places P with integer coefficients a_P, and $a_P \neq 0$ for only finitely many P. One aften writes $v_P(D)$ for the coefficient a_P, hence

$$D = \sum_P v_P(D)P.$$

The *support* of the divisor D is the finite set of places

$$\mathrm{Supp}(D) = \{P \mid v_P(D) \neq 0\},$$

and the *degree* of D is defined as

$$\deg(D) = \sum_P v_P(D) \deg P.$$

Let P be a place of F and x a nonzero element of F. The place P is called a *zero* of x if $v_P(x) > 0$ and a *pole* of x if $v_P(x) < 0$. The *zero divisor* of the element x is defined as

$$(x)_0 = \sum_{v_P(x)>0} v_P(x)P,$$

and the *pole divisor* of x is defined as

$$(x)_\infty = (x^{-1})_0 = - \sum_{v_P(x)<0} v_P(x)P.$$

The *principal divisor* of x is given by

$$\mathrm{div}(x) = (x)_0 - (x)_\infty.$$

All principal divisors have degree zero.

(4) With a divisor D of F one associates its *Riemann-Roch space*

$$\mathcal{L}(D) = \{x \in F^\times \mid \mathrm{div}(x) \geq -D\} \cup \{0\}.$$

This is a finite-dimensional vector space over K, its dimension is denoted by $\ell(D)$. One of the main results in the theory of function fields is the following theorem which gives a formula for the dimension of Riemann-Roch spaces:

(5) (*Riemann-Roch theorem*) Let F/K be a function field. Then there is a non-negative integer $g = g(F)$ such that:

(i) $\ell(D) \geq \deg(D) + 1 - g$ for all divisors D of F/K.
(ii) For all divisors D with $\deg D > 2g - 2$ we have $\ell(D) = \deg(D) + 1 - g$.

The integer g is uniquely determined by the conditions in (i) and (ii), and it is called the *genus* of the function field F. The rational function field $K(x)$ has genus $g(K(x)) = 0$.

(6) Let F/K and E/K be function fields with $F \subseteq E$. Then the extension E/F is a finite field extension. Let P be a place of F and let Q be a place of E. We say that Q *lies above* P (and write then $Q|P$), if the valuation ring of the place Q contains the valuation ring of P. We have the following facts:

(i) For all places P of F, the set of places Q of E which lie above P is finite and non-empty.

(ii) Let Q be a place of E. Then there exists exactly one place P of F such that $Q|P$, namely $P = Q \cap F$.

Now let $Q|P$ be places as above. Then there is a unique integer $e = e(Q|P) \geq 1$ such that $v_Q(z) = e \cdot v_P(z)$ for all elements $z \in F$. The number e is called the *ramification index* of $Q|P$. The place Q is said to be *ramified* over P if $e(Q|P) > 1$, otherwise $Q|P$ is *unramified*.

Also, there is an integer $f = f(Q|P) \geq 1$ such that $\deg(Q) = f(Q|P) \cdot \deg(P)$, and we call $f(Q|P)$ the *relative degree* of $Q|P$. The following formula ("*fundamental equality*") holds for any place P of the function field F:

$$\sum_{Q|P} e(Q|P) \cdot f(Q|P) = [E : F].$$

(7) Let F/K and E/K be function fields such that $E \supseteq F$ is a finite and *separable* extension. Then almost all (i.e., all but finitely many) places of E are unramified in E/F. Let P be a place of F and let Q be a place of E lying above P. Then one defines the *different exponent* $d(Q|P)$; this is a non-negative integer which has the following property :

$$d(Q|P) \geq e(Q|P) - 1 ,$$

with equality if and only if $e(Q|P)$ is not divisible by the characteristic of K. The divisor

$$\mathrm{Diff}(E/F) = \sum_{Q} d(Q|P)Q$$

is called the *different* of E/F. Note that the different of E/F is a divisor of the function field E/K, and we have $\mathrm{Diff}(E/F) \geq 0$. The support of $\mathrm{Diff}(E/F)$ contains exactly the places of E which are ramified in E/F.

(8) Consider again a separable extension E/F of function fields as in (7), and let P and Q be places of F and E with $Q|P$. We say that $Q|P$ is *wildly ramified* if the ramification index $e(Q|P)$ is divisible by the characteristic of K. Otherwise, $Q|P$ is said to be *tame*. By (7) we have that

$$d(Q|P) = e(Q|P) - 1, \ \text{if } Q|P \text{ is tame},$$

and

$$d(Q|P) \geq e(Q|P), \ \text{if } Q|P \text{ is wild}.$$

(9) Let E/F be a separable extension of function fields having the same constant field K. Then one has the following "Hurwitz genus formula" which relates the genera of F and E:

$$2g(E) - 2 = [E : F](2g(F) - 2) + \deg \text{Diff}(E/F).$$

This formula is crucial in order to determine the genus of a function field, since a function field F/K is often represented as a finite separable extension of a rational subfield $K(x)$. Then the Hurwitz genus formula becomes

$$2g(F) - 2 = -2[F : K(x)] + \deg \text{Diff}(F/K(x)).$$

About the Authors

Arnaldo Garcia; Instituto de Matématica Pura e Aplicada (IMPA), Rio de Janeiro RJ, Brazil; E-mail: garcia@impa.br

Cem Güneri; Faculty of Engineering and Natural Sciences, Sabancı University, İstanbul, Turkey; E-mail: guneri@sabanciuniv.edu

Ram Murty[1,2]; Department of Mathematics, Queen's University, Ontario, Canada ; E-mail: murty@mast.queensu.ca

Harald Niederreiter[3]; Department of Mathematics, National University of Singapore, Singapore; E-mail: nied@math.nus.edu.sg

Ferruh Özbudak[4]; Department of Mathematics, Middle East Technical University, Ankara, Turkey; E-mail: ozbudak@metu.edu.tr

Igor Shparlinski[1,5]; Department of Computing, Macquarie University, Sydney, Australia; E-mail: igor@ics.mq.edu.au

Henning Stichtenoth; Department of Mathematics, University of Duisburg-Essen, Essen, Germany and Faculty of Engineering and Natural Sciences, Sabancı University, İstanbul, Turkey; E-mail: stichtenoth@uni-essen.de and henning@sabanciuniv.edu

Alev Topuzoğlu; Faculty of Engineering and Natural Sciences, Sabancı University, İstanbul, Turkey; E-mail: alev@sabanciuniv.edu

---

[1] Thanks to Florian Luca and Francesco Pappalardi for a careful reading of the manuscript and many valuable comments.

[2] The author was supported in part by an NSERC grant.

[3] The author was supported by Australian Research Council Discovery Grants and by the MOE-ARF research grant R-146-000-066-112.

[4] The author was supported in part by the Turkish Academy of Sciences in the framework of Young Scientists Award Programme (F.Ö./TÜBA-GEBIP/2003-13).

[5] The author was supported in part by an ARC grant.

A. Garcia and H. Stichtenoth (eds.), Topics in Geometry, Coding Theory and Cryptography, 199–198.
© 2007 Springer.

Huaxiong Wang[3]; Department of Computing, Macquarie University, Sydney, Australia ; E-mail: hwang@ics.mq.edu.au

Arne Winterhof[6]; Johann Radon Institute for Computational and Applied Mathematics, Austrian Academy of Sciences, Linz, Austria; E-mail: arne.winterhof@oeaw.ac.at

Chaoping Xing[3]; Department of Mathematics, National University of Singapore, Singapore; E-mail: matxcp@nus.edu.sg

[6]The author was supported in part by Austrian Science Fund (FWF), grant S8313

Algebras and Applications